Everything You Know About Science Is Wrong

Everything You Know About Science Is Wrong

Matt Brown

BATSFORD

To my wife Heather. Our eyes first met across a crowded science publishing office...

First published in the United Kingdom in 2017 by
Batsford
1 Gower Street
London
WC1E 6HD
An imprint of Pavilion Books Company Ltd

ISBN: 9781849944021

A CIP catalogue record for this book is available from the British
Library.

20 19 18 17
10 9 8 7 6 5 4 3 2 1

Reproduction by Mission Productions, Hong Kong
Printed by 1010 Printing International Limited, China

This book can be ordered direct from the publisher
at the website www.pavilionbooks.com, or try your
local bookshop.

Contents

Introduction

Science is Fun! I remember reading that on the side of a pencil, given to me by a representative of the Royal Society of Chemistry. I was only 12 or so at the time, but that piece of stationery must have made its mark*. I went on to study chemistry at university and eventually found myself editing a chemistry journal.

And science really is fun, so long as you have a sense of curiosity. Didn't we all feel a thrill when we learnt that humans evolved from chimps, that glass is really a liquid and that nothing can travel faster than the speed of light? Facts like these give us a deeper insight into the world around us. We can impress our friends, or provide our children with convincing answers when they ask 'Why?'.

While science is fun, science mythbusting is even more fun. Despite the copious evidence from reality TV shows, not a single human being is descended from a chimpanzee. Glass most certainly is not a liquid, contrary to what you might have heard. And while the speed of light is fundamentally unbreakable in most circumstances, there are several sneaky ways to outpace it. In exploring such misconceptions, we gain a greater understanding of, and appreciation for, the real science behind the myth.

Mythbusting is not only fun, it's also important for everyday life. The world is full of pseudoscience – ideas that sound plausible and scientific, but are ultimately worthless. Whole industries are built on the credulity of a trusting public. Homeopathic medicine, detox diets, water ionizers and colonic irrigation all have the ring of scientific plausibility, yet none of them stand up to proper scrutiny. A good grounding in critical thinking can help us avoid wasting our time and money. Meanwhile, science is often twisted and misrepresented by politicians, campaign groups, newspaper columnists and others in positions of influence. With the looming spectres of climate change and antibiotic resistance, and game-changing technologies such as gene therapy and artificial intelligence, it is now more important than ever to be scientifically literate.

This book is a compendium of some of the most common misconceptions about science. Some widely held facts are just plain wrong. Others were once understood to be correct, but have since been overturned by new and better evidence. Still other facts are wrong under certain circumstances or don't present the full picture. It's true that the Moon orbits the Earth, but that's not the whole story.

Throughout the book I use the term 'science' in its broadest sense. You'll find sections that stray into the allied worlds of maths, engineering, medicine and technology. In many cases, especially the section on the nature of science, the brief entries barely scratch the surface. Whole bookshelves could be filled on the relationship between science and religion, for example. So, too, the many misconceptions about evolution, or dubious nutritional claims, which seem to spread as readily as acai berry jam.

In a book full of nitpicking and mythbusting, one does run the risk of sounding uppity; 'I think you'll find ...' are the four most annoying words in the English language. To avoid this, I hope I've kept the tone light and friendly throughout. For the same reason, I've minimised the references to scientific literature – the book is intended as a whimsical conversation starter rather than a fully annotated dissertation.

Finally, why should you believe my explanations over other sources? Excellent question. You shouldn't. Perhaps the greatest teaching of science is that we don't need belief in order to make sense of the world. Nothing should be taken at face value, including the entries in this book. The reader is encouraged to use what follows as a springboard. The realms of science are vast, fascinating and often misrepresented. Get digging, and let the nitpicking begin!

---

* FOOTNOTE The accompanying sticker, which punned 'Cuddle a chemist and see the reaction', proved less beneficial to my science career – but that's another story.

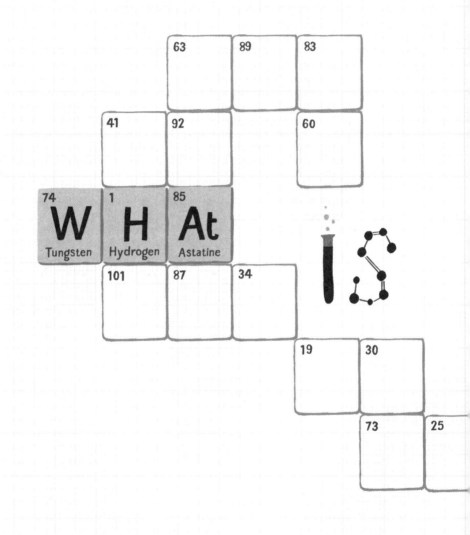

What is Science?

| 16 **S** Sulfur | 12 **C** Carbon | 53 **I** Iodine | **E** | 7 **N** Nitrogen | 58 **Ce** Cerium | 27 **?** |

12

116 68 2

62

45 54 40

9

Who do you picture when you hear the word 'scientist'? What do scientists do all day? Are they all superbrains?

All scientists look like this

In 2001, the first detailed guide to the human genome was published in the scientific journal *Nature*. This landmark study was written up by a truly international bunch. Dozens of scientists from the USA, UK, Japan, France, Germany, China, Ireland and Israel all collaborated on the genome. They represented 24 different universities and research organizations. This is positively parochial compared with a 2015 paper on the Higgs boson. Its remarkable tally of 5,154 authors worked in more than 50 countries. I've not met any of them personally, but I dare say not one of these scientists looks like the character drawn here.

Science is a broad church, employing people of every colour, creed, nationality, gender, temperament, hairline and body odour. You don't have to be crazy to work here, and it really wouldn't help. Yet all too often the word 'scientist' still conjures up the 'mad professor' stereotype. I saw one only this morning while watching an otherwise spellbinding pre-school show with my young daughter. Let us take a leaf from the notebook of Dr Frankenstein and dissect the stereotype scientist into his component parts.

Scientists have crazy hair The clichéd scientist never troubles a barber. He is either bald as an egg or else sports a frazzled, unkempt coiffure. The latter look was no doubt inspired by Albert Einstein, whose tousled white locks were more entangled than two electrons partaking in some spooky action at a distance*.

Big-haired boffins appear regularly on film and TV. Think Emmett Brown in *Back to the Future*, Gene Wilder's eponymous character in *Young Frankenstein*, or the goofy scientist in *Independence Day* played by Brent Spiner. Bald scientists are typified by such characters as Bunsen from the Muppets or Professor X from the X-Men. (The stereotype continues with meth-cooking chemistry teacher Walter White from *Breaking Bad*, though his hair loss has deeper meaning and one could hardly call him a cliché.)

I've known researchers with no hair, too much hair, wavy hair, spiky hair; blonde, brown, black, red and green hair, or all five at once. In 2012, following the successful landing on Mars of the Curiosity rover, mission engineer Bobak Ferdowsi was dubbed 'Mohawk Guy' for his crested locks. Particularly hirsute individuals can even join a Luxuriant Flowing Hair Club For Scientists. Researchers work at the cutting edge, in more ways than one. It hardly needs saying that, in reality, scientists sport as wide a range of hairstyles as any other sector of the workforce.

Scientists are male Do an image search for 'Solvay conference'. This is an occasional get-together of the world's top physicists, whose most notable gatherings took place in the early 20th century. A photo from the first conference, in 1911, shows two dozen or so leading scientists grouped around a table. Twenty-two of them have moustaches and one of them is Marie Curie. All are white. Sixteen years later, at the fifth conference, the photo shows 28 white men and Marie Curie. Not much had changed, save for a slight diminution in the popularity of moustaches.

---

* FOOTNOTE Just a little in-joke for those who know their quantum history.

Historically, science, medicine and technology were almost entirely the preserve of men, but it would be wrong to assume that women were totally excluded. The annual Ada Lovelace Day in mid-October celebrates the contribution played by women in these fields, both historically and today. Lovelace (1815–52) is often considered the world's first computer programmer; she devised algorithms for Charles Babbage's mechanical contraptions.

She and Curie are perhaps the most famous female scientists (using that term broadly), but there are plenty of other women who played an important role in the history of science. Caroline Herschel (1750–1848), for example, assisted with the discovery of Uranus and collated a catalogue of nebulae. Mary Anning (1799–1847) was among the greatest fossil hunters of all time, identifying some of the first plesiosaurs, ichthyosaurs and pterosaurs. Maria Mitchell (1818–89) was able to overcome the prejudices of her day to become a professor of astronomy at Vassar College, New York.

The 20th century saw still more eminent roles for women. In 1903, Marie Curie became the first woman to win a Nobel Prize (for Physics), for her work on radiation. She scooped the Chemistry prize eight years later for discovering polonium and radium. Her daughter, Irène Joliot-Curie, was the next woman to get a prize, taking the 1935 Chemistry award.

Such exceptional examples notwithstanding, the opportunities for women in science were dismal until relatively recent years. At the time of writing, only 17 women have received scientific Nobels; I lost count of the male winners somewhere around the 200 mark. Even today, the middle-aged man in the white coat remains the most common depiction of the scientist (try another image search if you want proof). What's the reality?

It's true that women still make up a disproportionately small part of the scientific workforce. Jaw-droppingly so. The United Nations Educational, Scientific and Cultural Organization (UNESCO) estimates that just 28.4 per cent of the world's researchers are female. This varies wildly by scientific discipline, and also from country to country. A handful of places, Latvia, for example, have more female than male researchers. Most nations are flimsy by comparison. The French scientific powerhouse is three-quarters male. Women are also less likely to be awarded a prize, invited to speak at a conference or chair a committee.

One common assumption is that this must be down to sexist hiring, but this isn't necessarily the case. A reputable 2015 study* looked into the recruitment of assistant professors at American campuses. It found that women were nearly always favoured over men with a 2:1 preference. And it didn't matter whether the person hiring was male or female. The results have been interpreted by some as 'positive discrimination', with recruiters under pressure to favour gender balance over other criteria.

So why do women remain under-represented in higher academia? The authors suggest that the problem is more on the 'supply side'. In other words, and for a complex variety of reasons, women are less likely to apply for senior roles. When they do, they are more likely to get an interview and secure the job than equally qualified male counterparts.

The picture is slowly improving. The number of men and women taking science to at least PhD level is now roughly equal in many Western countries, and the stats are getting better further up the career ladder.

Scientists are elderly Caucasians The stereotypical scientist is often depicted with white skin and even whiter hair. Stats are hard to come by, but it seems that reality tends towards the opposite. As with any hierarchy, the majority of scientific jobs are at the lower grades – the postdocs, technicians and junior researchers who do most of the experimental graft. Furthermore, this base is getting heftier. Scientific output has grown steadily over the past few decades, right across the globe. That implies ever-more jobs in science, most of which will be at the junior levels normally occupied by younger people. The elderly professor with unkempt white locks is very much in a minority.

How about ethnic background? In Western countries, as would be expected, the majority of the scientific workforce remains white, though disproportionately so. For example, 2010 figures show that 69 per cent of the US sector was white

---

* FOOTNOTE The paper is well worth a read. It's freely available online at http://www.pnas.org/content/112/17/5360.full

(compared with around 64 per cent of the wider population). If we look at the global picture, however, the Caucasian dominance is less clear-cut. Recent figures show that 40 per cent of the world's research and development now takes place in Asia. China alone accounts for 20 per cent of R&D. 49 per cent of all bachelor degrees in that country are in a scientific discipline. In fact, China produces more scientific graduates than any other country. With India also on the rise, and Japan and Korea maintaining a strong research base, perhaps the stereotype scientist will soon be a young Asian.

Scientists always wear a lab coat and safety specs: OK, it's undeniable. Some scientists do wear white coats and eye protection. Anyone who handles hazardous substances or anything that might stain their civvies will don a lab coat, while plastic glasses are mandatory in any chemistry laboratory. Scientific couture is much broader than you might imagine, however. A marine biologist would find her underwater progress impeded by a flapping white coat. A field geologist would feel the wind chill. Many theoretical physicists do their best work at home – anyone who wears safety equipment in bed is probably engaging in pursuits other than the scientific. Some scientists wear suits, many go to work in everyday casual dress and, yes, one or two have corduroy jackets with brown elbow patches. To assume that a scientist must be dressed in lab gear is like supposing that all soldiers wear bearskin hats.

Scientists play with bubbling test tubes You're not a proper movie scientist unless you surround yourself with coloured potions that churn and foam from exotic glassware. Researchers in the real world *never* decorate their workplace with such frenzied apparatus. For starters, most scientists are not chemists and so have little business swilling around purple liquids or dodging clouds of smoke. Even those who do muck about with reagents would find such scenarios bizarre. Anything that fizzes, smokes, effervesces or otherwise expels material must be contained within a ventilated fume cabinet. Brightly coloured liquids are, broadly speaking, the preserve of undergraduate teaching labs. Having lived through a degree in chemistry, and another in biochemistry, the only time I ever saw a foaming test tube was when we filched some glassware for an experimental cocktail party.

Another gimmick you'll often spot, even on serious scientific documentaries, is the preponderance of purple lighting in laboratories. Doesn't happen, except in the rare labs that use UV lamps for visualizing certain reagents. The mauve hues seen on TV are purely for visual interest. Laboratories are often brightly lit, with stark, white surfaces – great for observational research, but a bit severe for television interviews.

You have to be really clever to understand science

Scene: a suburban dinner party. Friends of the hosts are chatting, getting to know one another.

'So, what do you do for a living?'

'I'm a scientist. I work at the university.'

'Huh. Wow. I was never clever enough to do science. All that maths and stuff. You must be really brainy.'

'Huh, well, not really. I pretty much just count fruit flies for a living. Why, what do you do?'

'Oh I'm a banker.'

'Really? All those numbers and percentage rates? You must be *really* clever.'

Anyone who's worked in science has had a similar conversation. That 'S' word. It fuddles people. Try saying 'I chew glass for a living' or 'I'm an interpretive dancer who works with bees' and you'll get a similar look: minor amazement with a hint of unease. Perhaps it has something to do with the way scientists are portrayed in the media. To take three recent headlines:

'Clouds don't have silver linings, say global warming boffins'

'"Vaping" can cause lung damage, say Manchester boffins'

'Glum, depressed ... and addicted to Facebook, Twitter? There's a link, say medical eggheads'

All use language that distances the reader from the scientists. Terms like boffin and egghead can be thrown in an affectionate way, but usually the context is mildly derogatory and anti-intellectual. 'Global warming boffins' does a double disservice. It adds to the impression that you must be super-intelligent to study the climate, while the dismissive tone of 'boffin' suggests a group of people you wouldn't want to be part of, as though it's uncool to be intelligent.

People think that science is hard. People think science is not for them. People think that science is only for 'boffins' and 'eggheads'. Yet science is born of the most basic of human instincts: curiosity about the world. Young children never stop asking 'Why?' and 'Why not?'. Some would say the instinct gets beaten out of us as we grow older; that classroom teaching turns people off science. But does it? Everybody would click the headline 'Life discovered on Mars'. Nobody would be indifferent to a drug that could cure dementia or double intelligence. Even esoteric concepts such as gravitational waves and the Higgs boson generate enormous public interest. You only have to look at the rise of science festivals and 'geek culture' to see that people *are* interested in science – they just don't necessarily want to be the ones *doing* the science.

Perhaps it takes a certain kind of person. Many children find science classes difficult, to the point of capitulation. On the other hand, those of us who enjoyed science sometimes struggled with humanities. With science, it's easier to know what's expected of you. If an exam question asks 'How many electrons in an atom of carbon', there's really only one correct answer. If an English literature exam asks you to discuss the motivations behind Macbeth's slaying of Duncan well, where do you start? Science, at least at the junior classroom level, is underpinned by straightforward rules, while humanities requires more interpretation and subjectivity. Generally speaking.

Of course, science gets more specialized as we progress through education. A student learning the biology of a cell will have to memorize and understand a whole new vocabulary. Terms like anaphase, meiosis, endoplasmic reticulum, ADP and – personal favourite – Golgi apparatus can be off-putting. Then again, millions of cricket fans can cope with phrases like maiden overs, deep backward square legs, sticky wickets, LBWs and googlies.

And what about the maths? Many people are put off the sciences because of a perception that it's all about formulae and equations. It's true, most scientific disciplines do require a certain amount of number crunching. But then so does accountancy, banking, plumbing, photography, video game design and a host of other professions.

Like most things in life, success at science isn't so much about being smart, as having a passion for the subject and putting in the work. As with cricket, a bit of effort to learn the rules and terminology pays off with a lifetime of enjoyment.

Researchers always follow the scientific method

In a perfect world, all research would be conducted according to the scientific method. It's a bit like a recipe, and goes something like this:

- **Ask a question**, like 'Why isn't this pan of pasta bubbling?'.
- **Formulate a hypothesis** 'A watched pot never boils'.
- **Test your hypothesis with experiments** Measure the time taken for a pot of pasta to heat to boiling point. Is it quicker when you look away?
- **Collect enough data** to be sure you've been thorough – you'll need to time the pot on dozens of occasions (both while watching and not watching) to be sure your data sets are large enough to draw a conclusion. A good scientist would also try the experiment with different pots, different levels of water, various durations of watching, and with different foodstuffs in the water, changing one variable at a time while keeping stove temperature and pressure constant.
- **Draw conclusions** No significant difference could be found. The time taken for a pot of water to boil is independent of whether it is observed.

In practice, science rarely runs according to the scientific method. Research is dynamic, evolving and messy. Our culinary scientist might quickly abandon the research when he figures out that the hypothesis is clearly ludicrous. Perhaps she'll spot a side effect that is much more interesting to study, such as the taste and texture of pasta cooked for different time periods. He might hear from a rival laboratory that all the grant money is going into sauce research, and shift his focus accordingly. Or she might have no hypothesis at all, and simply want to find out what happens when you boil pasta for 24 hours in orange juice. The point is that scientists make progress as much by hunch, whim and serendipity as by following the commandments of the scientific method.

That's not to say their work is sloppy or haphazard. A good scientist might set out on a study through intuition or even desperation, but he will still use all the proper controls, checks, balances and repetitions to make sure that measurements are statistically sound.

Further, not all realms of science use experimentation. How, for example, can we ever know what goes on inside a black hole? Nothing can escape from these enigmatic objects. We can't venture inside the black hole and make measurements to test our hypothesis – or if we could, we wouldn't be able to tell anyone about it because nothing we could say, do or transmit could make it out of the black hole again. Many other areas of science hit barriers to experimentation. How can we test the hypothesis that there are multiple universes? What happens to particles that collide at energies hundreds of times those possible in the Large Hadron Collider?

Ideas like these are untestable, at least with current technology. That makes some people uncomfortable. A proper scientific theory should be 'falsifiable'. It should be open to testing by observation. If I were to speculate that the Moon is made of cheese, you could prove my hypothesis wrong by going to the Moon, taking a sample and baking it into a lasagne. It is a falsifiable theory. Were you to tell me that rubbing goat dung onto my scalp is a certain cure for baldness, I could test your claims by applying the substance to my pate to see what happens. One test is never enough, so I might then organize a clinical trial involving lots of bald participants and lots of goat dung. With enough participants I could find a statistically significant result to show whether goat poo is or is not an effective cure for baldness. The theory is again falsifiable. Now suppose I came up with the theory that black holes are gateways to the afterlife. Enter a black hole and you'll find yourself in glorious communion with the souls of your departed loved ones. How could we begin to confirm or refute such a hypothesis? We can't. It is non-falsifiable, and therefore non-scientific.

My example above sounds ludicrous, but some firmly established theories are equally hard to falsify. Take string theory, for example. It posits that reality is built up from tiny strings that occupy multiple dimensions (see a later chapter for more on that tangle). String theory is as close as we've yet come to bringing the very small (the quantum world) and the very large (the cosmos) into the same set of equations, and thus arriving at a 'theory of everything'. The trouble is, it is almost impossible to build an experiment that could prove or falsify all of string theory's predictions. The energies involved are way beyond any experiment we might conceive in the foreseeable future. That said, 100 years ago we had no way of being sure the Moon was not made of cheese.

Even down-to-earth sciences like geology rarely follow the scientific method. A geologist working out in the field might collect rock and soil samples to learn more about the terrain and its history. At no point is she performing an experiment. This is simply the collection of evidence, but it is still a form of science.

All this is to say that the human endeavour known as science often extends beyond the standard recipe of the scientific method – a set of steps dreamt up by historians of science rather than scientists themselves. Some results are found by serendipity, others by the merging together of different experiments. Some areas of science dispense with experiments and data altogether, and are purely theoretical, or else reliant on a computer model of reality. Perhaps the best science of all happens when someone says 'I wonder...?'.

We don't need scientists to tell us how the world works, we can just use common sense

The world is a very strange place. Common sense tells us that the Sun rises in the east and sets in the west. More detailed observation shows that the Sun doesn't move at all (at least not in the way we think it does). The Earth's rotation makes the Sun appear to arc across the sky.

Common sense would have us shouting foul play if a 50-ball lottery draw revealed the numbers 1, 2, 3, 4, 5 and 6. Simple statistics shows that this is as likely as any other draw.

Common sense, for some reason, makes us think that heavy items fall more quickly than lighter ones. Experiments conducted centuries ago proved this to be untrue.

The world doesn't always match our intuitions. Even making a cup of tea can be fraught with misconceptions. My friend Geoff recently noted (with some anguish) that office packs of his favourite blend come in portions of 480 bags. 'Why can't they do it in packs of 500?', he wondered. Five hundred is, of course, a more rounded, simple number. Only it isn't. Humans like to see stuff falling into multiples of ten, perhaps because we have ten digits on our hands. Hence, 500 feels like a good round number whereas 480 feels like, well, a mistake.

Look at it dispassionately, though, and 480 has much more going for it: 480 has 24 factors – that is, whole numbers by which it can be divided without remainder. For the record, these are 1, 2, 3, 4, 5, 6, 8, 10, 12, 15, 16, 20, 24, 30, 32, 40, 48, 60, 80, 96, 120, 160, 240 and 480. The seemingly landmark number of 500, by contrast, only has 12 factors (1, 2, 4, 5, 10, 20, 25, 50, 100, 125, 250 and 500). It runs contrary to our expectation, but 480 is a more useful number than 500 – particularly if you're organizing lots of tea parties and you want to maximize the number of ways you can share out the teabags. We've all been in that situation. This is partly why old-fashioned systems of 12 inches to 1 foot and 24 hours in a day still hold sway. Twelve and 24 can be divided more readily than ten and 20, making mental calculations more convenient.

Science and mathematics are the best tools we have for weeding out our human biases. This is particularly true when it comes to the quantum world. At the unfathomably tiny scales of atoms and subatomic particles, all kinds of weird stuff goes on that would be the very antithesis of common sense. Particles can pop into being from nowhere, and then disappear as mysteriously as they arrived. Others can exist in two opposite states at the same time. Nobody could have conceived any of this through common sense, yet solid theoretical work and countless experiments have shown it to be true.

So-called common sense is informed by an accumulation of everyday experiences. To paraphrase Einstein, it is merely a collection of prejudices acquired in our formative years. Humans cannot directly experience the quantum world, and nor can we truly grasp the vast expanse of the cosmos (let alone additional dimensions and realities). Without these experiences to feed it, common sense can tell us very little about the wider reality.

Scientists get everything right

One of the most referenced scientific papers of recent times dealt not with the fundamental forces of nature or the inner workings of the cell, but with the way scientists go about their business. The 2005 article, *'Why Most Published Research Findings are False'* by John Ioannidis caused, as you might expect from its title, a certain amount of kerfuffle. The problem was further highlighted in 2015, when a paper in *Science* described efforts to re-run 100 psychology experiments previously published in reputable journals. Only about one-third could be replicated, casting grave doubts on the findings of the other two-thirds. How could this happen?

The idea that large swathes of published research might be wrong is deeply troubling. Publication drives science. Part of the deal with being a scientist is that you need to write up your experiments and then seek to publish them in a scholarly journal. A scientist's reputation depends, in large part, on how often he or she publishes, and in which journals. Pressure mounts on the scientist to get their work into these journals. Publishers, meanwhile, are under commercial pressure to launch new journals and pump out more papers, in order to grow their business. Neither of these effects does much to encourage quality.

Any reputable journal will use a process known as peer review to help spot errors in a scientific paper it is considering for publication. The paper is sent off to several of the author's peers – people working in the same field of science – who usually remain anonymous from the author. These peers scrutinize the work and report back to the editor, often suggesting changes to the manuscript, or additional experiments they feel are necessary to support the author's conclusions. The editor then works with the author to see that the peer reviewers' comments are taken into account. Once everyone is happy, the paper is published and becomes part of the scientific record.

Peer review can help weed out some of the bad science, but not all. Scientists can fail in all the ways that other humans can. They might subconsciously massage their data, giving more weight to points that support their ideas while assuming that wild results must be errors. They might be poor at statistics and not realize that their sample size is too small or that the measured effect is too tiny to be significant. Some, hopefully a very small number, deliberately skew results in order to push an agenda and further their careers.

Such flaws can be tricky to spot. For example, authors rarely supply their raw data. By the time the write-up reaches an editor, the data have been neatly packaged into graphs and tables, and placed in the wider narrative of the paper. It's difficult for peer reviewers to spot if the author has taken liberties when choosing what to include. Even when peer reviewers can look at the untampered raw data, they may not have the skills or inclination (peer review is usually unpaid) to wade through and repeat the statistical analysis. Taking all these effects together, it's small wonder that errors make it into publication.

Does this mean that science is broken and can't be trusted? No. The system is self-correcting. If results have been tweaked or faked they will ultimately be uncovered when another scientist seeks to build on the work. Most of the science that filters down to the wider public has been thoroughly scrutinized and reproduced by other scientists. An important scientific consensus, such as the evidence for man-made climate change, will have been poked and probed from every conceivable angle.

Even so, efforts are now under way to reduce the amount of bad science that gets published. Initiatives such as the Open Science Framework encourage authors to register their studies before they undertake them, and to publish full data sets for anyone to scrutinize.

Science and religion are always opposed

Both religion and science seek to explain our origins and our place in the Universe. They can be seen as rival systems of understanding, one based on belief and superstition, the other on measurement, observation and evidence. So different are these systems that many have argued that they cannot possibly be compatible. Either religion goes or science goes.

The unavoidable example of this clash is Galileo. Everyone knows that the astronomer was locked up in the 17th century for supporting the view that the Earth might revolve around the Sun. 'Vehemently suspect of heresy,' was the charge. Galileo was able to avoid the death penalty by renouncing his opinions, but he remained under house arrest for the rest of his life. He was neither the first nor the last rational thinker to be persecuted for challenging a supernatural view of the Universe.

To leave things there would be to present an over-simplistic picture. Galileo remained, as far as anyone can know, a devout Catholic even after his run-in with the church authorities. He built upon the astronomical work of Nicolaus Copernicus, who was not only religious, but also a Catholic priest. It was an age where, at least in the Western world, almost everybody was a believer. Those who harboured doubts about God would usually keep them quiet, or face alienation and persecution. Needless to say, most scientists up until the 20th century were at least outwardly religious.

Chief elder of all religious scientists was Isaac Newton. The man who untangled the heavens, devised the calculus and picked apart the rainbow was also a deep spiritual thinker. His religious writings are more abundant than his scientific. Newton was raised as an Anglican but – ever the freethinker – would later tend towards the more peculiar flavours of the faith. In private, he rejected Christ as part of the Holy Trinity, and believed that worshipping the Messiah was idolatry. He kept this secret. Even a man of Newton's clout wasn't immune to accusations of heresy.

Another key scientist with a strong religious faith was Gregor Mendel. Mendel is often called the father of modern genetics for working out the rules of heredity in plants – that is, how certain traits such as plant length, colour and shape are passed down from one generation to the next. His work was published in 1866 but remained obscure until the 20th century, when scientists began to gain a better understanding of genetics. When not busy founding one of the cornerstones of modern science (and also publishing widely on meteorology), Mendel was also a religious man. A very religious man. Search for a photo and you'll see that the most commonly displayed image shows Mendel sporting a crucifix the size of a sandwich. That's because he served as an Augustinian friar, and later became an abbot. For Mendel, as for so many others, there was no conflict between serving God and exploring the patterns of nature.

Galileo, Newton and Mendel were not alone as scientific men of faith. The list is long and includes many famous names: Robert Boyle, Johannes Kepler, Gottfried Leibniz, Joseph Priestley, Alessandro Volta, Michael Faraday, James Clerk Maxwell and Lord Kelvin, to name but a few. Roger Bacon, a 13th-century proponent of the scientific investigation of nature, was a Franciscan friar. William Whewell, the man who coined the word 'scientist' in 1833, was also a Christian theologian. So strong is the crossover that Wikipedia maintains a list of religious scientists. It currently stands at over 200 names – and that's just Christians worthy of biography. Islam, too, has a long history of supporting science. We owe an incalculable debt to Islamic scholars of the Middle Ages for preserving the writings of the Classical world. Wikipedia also houses a list of Islamic scientists. For what it's worth, this is even longer than the Christian list. And much could be written on scientists of other faiths.

The point is, science and religion have a long history of peaceful coexistence, occasionally punctuated by incidents like the gagging of Galileo (itself often exaggerated). They need not stand in opposition. Many scientists today hold religious or at least spiritual views, and draw on their beliefs for creative inspiration. You could hardly find a more prominent example than Francis

Collins. He currently heads up the National Institutes of Health in the USA, having previously led the Human Genome Project. This most eminent of scientists is also a devout Christian and often speaks of compatibility between the worldviews. Conversely, the Vatican has long supported its own Pontifical Academy of Sciences (of which Collins is a member). I once had the privilege of meeting the 'Pope's Astronomer', Brother Guy Consolmagno, who now heads up the Vatican Observatory and is a leading meteorite researcher. One could produce many more examples, for the links between science and religion are not as severed as some might claim.

After all, some of the most fundamental questions still remain beyond the grasp of science. What is human consciousness? How did life begin? Why is anything here at all? Scientists have ideas, but not yet answers.

To infinity and beyond

Into the stratosphere and on to the stars.

The Wright brothers performed the world's first heavier-than-air flight

To paraphrase Douglas Adams, flying is the art of throwing yourself at the ground and missing. For most of our existence as a species, humans have been inept at this art. The most noted aeronaut of antiquity – the mythical Icarus – is famous for crashing.

It wasn't until the late 18th century and the advent of the first balloons* that humans took to the air in any competent way. Temperamental and at the mercy of the winds, balloons remained our only chariots of the sky until 1903, when the Wright brothers made the first winged flight. That's the commonly believed story, anyhow. In fact, the world's first recorded heavier-than-air flight took place long before either Wright brother was born. The credit goes to a Yorkshireman who was rapidly approaching his eightieth birthday.

Sir George Cayley had lived a long, distinguished career. This ennobled polymath had invented self-righting lifeboats, seat belts, an early combustion engine and the spoked wheels that we still use on bicycles. He'd also made time to serve as a Member of Parliament and to found the Royal Polytechnic Institution, now the University of Westminster.

Cayley had saved his most impressive accomplishment for his twilight years: the first recorded flight in a winged aircraft. He'd been building up to the feat all his life. His teenage notebooks of 60 years before include sketches of credible flying machines. Soon after, he'd described the four forces that would affect a body in flight: thrust, drag, gravity and lift. In 1853, everything was to come together in a field at Brompton Dale, near Scarborough, England.

Cayley had built a glider large enough to seat an adult. A wooden cockpit, resembling a rowing boat on wheels, was slung beneath a canvas wing, all stabilized by twin tails. As one newspaper later described it, 'a sort of light vessel, with the requisite appendages to agitate and float upon the air, like the flying of a bird'.

---

* FOOTNOTE: Another misconception, here. It's often said that the Montgolfier brothers made the first human balloon flight. They didn't. Not really. One of them, Étienne, did become the first human to rise in the air, on or around 15 October 1783. His balloon was tethered close to the ground, so it was more of a taster experience than a proper flight. The glory instead goes to Pilâtre de Rozier and Marquis François d'Arlandes. Their flight on 21 November 1783 in a Montgolfier contraption covered around 8 km (5 miles) across Paris. It was a magical time. Just ten days after the first hot-air balloon, Jacques Charles and Nicolas-Louis Robert completed the world's first helium balloon flight, also in Paris. It's always struck me as remarkable that, after thousands of years dreaming about it, humans found two different ways to fly within the space of a fortnight. As far as is known, the Montgolfiers never once made a true balloon flight, even though their name is so intimately associated with the technology.

The aged Cayley was too frail to fly the craft himself. It is thought that his coachman John Appleby was at the controls, but we may never know the identity of that first pilot for certain.

We do know that the vehicle lifted off from a hilltop and carried its rider some 150 m (492 ft) before hitting dirt. Cayley had launched earlier model craft, and possibly even an earlier glider with a child pilot (now there's a Disney film if ever there was one), but this short hop over Brompton Dale was the first properly documented winged flight anywhere in the world.

Many other glider pilots would take to the skies over the following decades, though the technology remained precarious. The Wright brothers were the first to successfully incorporate an engine, after hundreds of preparatory flights in gliders. They made the history books on 17 December 1903 with a succession of powered flights at Kitty Hawk in North Carolina. The world would never be as large again.

Sputnik was the first artificial object in space

Sixty years ago, a ball of metal the size of a toilet bowl changed the world. Sputnik 1, launched by the Soviet Union on 4 October 1957, circled the globe for three months, beaming telemetry down to anybody with a shortwave radio. Never in human history had a simple series of beeps provoked so many emotions. Some feared for a future in which a totalitarian state could command the planet's skies. Others were in awe at the pace of technology; it had been little more than 50 years since the first powered flight of the Wright brothers. The achievement also acted as a fillip for other nations, most notably the USA, to speed up their own space programmes. Sputnik proved a tiny but powerful moon.

It was not, however, the first bit of kit to enter space. Hundreds of earlier rockets had shot well beyond the boundary, only to fall immediately back to Earth on a ballistic trajectory. Sputnik went much further in having enough speed to achieve an orbit and stay aloft, but it would be wrong to suggest that its launch in 1957 was humanity's first toehold in space. That landmark had occurred some 15 years earlier, during the throes of the Second World War.

The Nazi regime invested heavily in rocketry, a potentially game-changing technology that could deliver a devastating explosive to a target hundreds of miles away without risking an aircrew. A large test facility was built in Peenemünde on the Baltic coast. After a number of false starts, an A-4 liquid-fuelled rocket achieved an altitude of 85 km (53 miles) on 3 October 1942. 'We have invaded space with our rocket,' said programme leader Walter Dornberger, 'and for the first time – mark this well – have used space as a bridge between two points on the Earth; we have proved rocket propulsion practicable for space travel'.

That first flight only scraped the edge of space. The boundary is not universally agreed; one definition puts it at 100 km (62 miles), while US astronauts earn their wings at just 80 km (50 miles) above mean sea level. However, later rocket flights during the war reached a maximum altitude of 189 km (117½ miles), well beyond most definitions of the threshold. The Germans, then, were the first to launch objects into space. They did so hundreds of times, and 15 years before Sputnik.

Although many of those early rocket engineers dreamed of space exploration, the A-4 was immediately put into service as the V-2 weapon of war. Hundreds would be fired on cities such as London and Antwerp, killing as many as 10,000 people. That number was overshadowed by the estimated 25,000 deaths in the German labour camps that built the rockets. It was not an auspicious start for the conquest of space.

Once the war finished, many of the Nazis' engineers were shepherded off to the USA and Soviet Union to assist those nations in developing their own rocket technology. Wernher von Braun, who had been instrumental in the creation of the V-2, went to America, where he would eventually inspire the Saturn V Moon rockets. The Soviets, meanwhile, bagged Peenemünde and other V-2 sites, as well as a number of rocket scientists. Under Sergei Korolev, the Soviets improved upon the German technology, eventually allowing the launch of Sputnik 1 and the human space programme.

The Great Wall of China is the only man-made object visible from the Moon

This well-known nugget dates back much further than you might imagine. Writing about Hadrian's Wall in 1754, the English antiquarian William Stukeley observed:

'This mighty wall of four score miles [129 km] in length is only exceeded by the Chinese wall, which makes a considerable figure upon the terrestrial globe, and may be discerned at the Moon.'

It was pure speculation, of course, coming over 200 years before humans went anywhere near the Moon. But, as with so many false facts, it has that ring of plausibility. Everyone knows the Great Wall of China is a preposterously impressive structure – it must surely be visible from the Moon?

Well, it's not. Not without a telescope, anyway. The Great Wall of China is, on average, around 6 m (19⅔ ft) wide. The Moon is, on average, 370,139 km (230,000 miles) away from the Great Wall of China. Think about those numbers for a second and you soon realize that the landmark is nowhere near thick enough to be seen from the Moon. Standing on the lunar surface, you'd do well to make out China, never mind its Great Wall. Indeed, no man-made object on Earth can be seen from the Moon. Even the night-time glow of a big city would be beyond the limits of the human eye at such a distance.

A watered down version of the myth holds that the Great Wall is the only man-made object visible from space – usually taken to mean low Earth orbit a few hundred kilometres up. There are two problems with this claim. First, the edifice is tricky if not impossible to make out even from this distance. China's first man in space, Yang Liwei, tried his best to spot the national monument without success, though others reckon to have spied it under ideal circumstances. Similarly, the pyramids of Giza are invisible to the naked eye from low Earth orbit, as confirmed in a tweet by astronaut Tim Peake.

A second flaw with the myth is that many other man-made structures can be seen from orbit. For starters, any major city shines out like a beacon during periods of darkness, and many can be identified during daylight too. Large dams are easy to spot, thanks to the unnatural effect they have on rivers. Long, straight roads of contrasting colour to the surrounding countryside are also visible. The patchwork-quilt effect of agriculture is another example. It's even possible to see a man-made Earth from space. A network of 300 artificial islands off the coast of Dubai has been sculpted from sand to resemble the continents. Mankind's handiwork is also evident at sea. The oceans now contain vast nautical acreages of algal bloom, which feed on the waste matter we pump into the sea.

Astronauts float in zero gravity

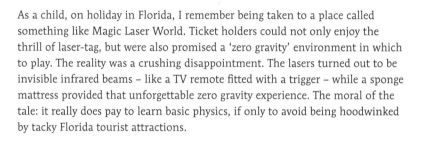

As a child, on holiday in Florida, I remember being taken to a place called something like Magic Laser World. Ticket holders could not only enjoy the thrill of laser-tag, but were also promised a 'zero gravity' environment in which to play. The reality was a crushing disappointment. The lasers turned out to be invisible infrared beams – like a TV remote fitted with a trigger – while a sponge mattress provided that unforgettable zero gravity experience. The moral of the tale: it really does pay to learn basic physics, if only to avoid being hoodwinked by tacky Florida tourist attractions.

Astronauts in Earth orbit are also cheated of a true zero G experience. Their surroundings might be a little more convincing than the bouncy furnishings of Magic Laser World, yet they are still very much beholden to the force of gravity. Those on board the space station are tugged by the Earth with around 89 per cent of the force felt by those of us stuck on the surface. So how come everyone assumes they're in zero gravity?

Imagine you're in a plummeting lift. You'd find yourself in free fall, able to bounce around the interior for the last few seconds of your existence. It would feel like zero gravity, even though your imminent appointment with the bottom of the lift shaft would soon reveal otherwise. It's the same on the space station. That magnificent craft is falling towards the Earth, like a ball dropped from a tower. It just so happens to be moving sideways at such a speed that it keeps missing the Earth. Effectively, the space station is falling around the curve of the planet. If you're new to this idea, stop and have a think about it, rather than moving on. It's one of those concepts that seems baffling at first, but once you've figured it out, you'll feel proper clever.

So the astronauts on board are in free fall, but it's a plummet without end. The effect is a feeling of weightlessness, as though free from the force of gravity. The big old Earth is still there, pulling at almost 90 per cent of its surface attraction, but as everything around you is falling at the same speed, you don't notice the tug. It feels like zero G, even though it's not.

As we move further from the Earth, the pull of gravity gets weaker and weaker according to what's called the inverse square law. If you increase your distance by a factor of three, the gravitational attraction will decrease by a factor of nine (that is, three squared). You can never entirely escape the clutches of gravity. Anything that has a mass is attracted by gravity to every other thing with mass. Your Great Aunt Jean feels the gravitational pull of your neighbour's lawnmower, which itself feels the tug of the Orion Nebula. The forces in these examples are imperceptibly weak, but present nonetheless. Even an astronaut who has fled the Solar System into interstellar space, and who finds herself billions of miles from any planet or star, is still unable to enjoy true zero gravity. She is still party to the gravitational tugs of everything in the observable universe. These forces, from every direction, will more or less balance out, and they are so weak that she wouldn't feel a thing. The claims of Magic Laser World notwithstanding, we can never be truly free of gravity.

Without a heat shield, spacecraft re-entering Earth's atmosphere would burn up from friction

Anyone who's ever belly-flopped into a swimming pool will appreciate that moving quickly from an area of low density to one of higher density causes something of a shock. A similar thing happens to a homecoming spaceship. Moving from the near vacuum of space to the gaseous cushion of the upper atmosphere can be a stressful ride, and the vehicle needs protective shielding to prevent a catastrophe.

The effect is often blamed on friction. It makes sense when you think about it. The vehicle is moving at colossal speeds, typically 27,000 km/h (17,000 mph) from the space station, or 40,000 km/h (25,000 mph) from the Moon. The upper atmosphere enjoys only a light dusting of molecules, but smack into them at speeds like this, and your belly will smart from the friction.

Only, it doesn't quite work like that. The heat is mostly caused by compression, not friction. Spacecraft tend to be bell-shaped, or else have a flat surface like the underside of the space shuttle, which leads the way during re-entry. As it bolts into the atmosphere, the spacecraft's flat side compresses the gases directly in front of it. The air is hit at such speed that it hasn't time to move aside and instead piles up in front of the vehicle. Basic physics tells us that gases under compression heat up – think how your bike pump warms as you squeeze the air inside it. This bubble of superheated air does not directly touch the vehicle thanks to a shockwave that separates the two. Significant heat is still transferred to the shield by radiation. Without the shield, this heat transfer would burn up the vehicle, but friction would play little part.

Incidentally, orbiting spacecraft don't *have* to re-enter the atmosphere at hypersonic speeds. It would be quite possible to reverse speed while still in space, and then gently pass through the upper atmosphere without the need for a heat shield. To do this would require a swift and powerful deceleration, essentially performing the launch sequence in reverse. Such a manoeuvre would need tonnes of additional rocket fuel. It's far more practical to use the atmosphere to slow down, rather than building a much bigger rocket to carry the deceleration fuel.

That said, some objects do come down from space, or near space, without a shield. Felix Baumgartner, who famously leaped from a balloon 38.6 km (24 miles) above the Earth in 2012, survived in just a spacesuit. How? He wasn't in orbit and, therefore didn't need to decelerate from those colossal speeds. Likewise, ballistic missiles travel into space before falling onto their unfortunate targets, but never achieve the great speeds needed for orbit. Their need for thermal protection is therefore much reduced.

The seasons are caused by the Earth getting closer to and farther from the Sun

Winter is coming. The Earth circles the Sun in an elliptical orbit, which by definition means it strays further from the orb during certain parts of its cycle. Like a dance around a campfire, the heat increases as we move in, and slips away as we move out. Ergo, the summer and the winter.

It's a seductively simple explanation, isn't it? A moment's ponder is enough to overturn the idea, however. If winter is caused by the Earth moving away from the Sun, then we'd expect everywhere on the planet to decrease in temperature at the same time. Yet we're all familiar with the idea of Australians basking in Christmas sunshine, while Canadians hunker down against the snows. When it's winter in the northern hemisphere, it's summer to the south; and vice versa. Likewise, those living at the equator experience little change in temperature throughout the year – although many places have a 'wet season' and a 'dry season'.

It's true that the Earth's distance from the Sun does vary, but not by much, and the effect on temperatures is minute. Instead, seasons are all about tilt. Imagine a giant rod passing through the Earth from pole to pole – this is the planet's axis, around which it revolves. That rod does not point straight up and down with respect to the planet's orbit. Rather, it is tilted at 23.5 degrees – about the angle you might attempt with a ladder against a wall. This means that different parts of the Earth receive differing amounts of energy from the Sun. The hemisphere mostly pointing toward the Sun basks in more direct sunlight; while the parts of the Earth pointing away get less, and at a more oblique angle. If you've ever tried to start a fire with a magnifying glass, you'll know that angled sunlight isn't as warming as more direct rays.

As the planet progresses around its orbit, the portion tilted towards the Sun gradually shifts. In other words, we get a change of season.

A related misconception is to assume that all four seasons are the same length. Seasons are defined by the equinoxes and solstices as Earth makes its progress around the Sun. Winter lasts around 89 days, spring 93 days, summer 94 days, and autumn 90 days. Winter is the shortest season.

Moons circle planets, and planets circle stars

It all looks very straightforward on paper. The eight planets of the Solar System revolve around the Sun. All these planets save Mercury and Venus are in turn orbited by moons. And so the wheels of the cosmos turn, like cogs in a watch. By now, you won't be surprised to hear that it's a little more complicated than that.

Recall that every object has its own gravitational field. The Moon is held in orbit by its gravitational attraction to the Earth. But the feeling is mutual. Earth also feels a pull from the Moon. Imagine a tug-of-war between strongman cartoon character Popeye and feeble bear of very little brain Winnie-the-Pooh. Spinach or no, we'd expect Popeye to win every time, but he would still feel some pull from his sleepy ursine opponent. So too with the Earth and the Moon. The planet pulls stronger, but its satellite is not without its own counter-heave. By the same token, the Moon also feels the gravitational pull of the Sun. In fact, the Moon feels a greater attraction to the Sun than it does to the Earth – more than twice as much. If our planet were to disappear tomorrow, the Moon would carry on happily orbiting the Sun, just like a planet*.

---

* FOOTNOTE Even so, it would be a bit much to say that the Sun sometimes goes round the Moon, as claimed by Vanessa Williams in the 1992 chart-topping song 'Save The Best For Last'.

Indeed, Earth's moon is something of a whopper and has some claims to planethood in its own right. Our rocky satellite is, for example, much larger than Pluto, which was considered a planet until 2006. It is the fifth largest moon in the Solar System, behind Jupiter's Ganymede, Callisto and Io, and Saturn's moon Titan. However, in relation to the size of the planet it orbits, our moon is way more impressive than these satellites of the gas giants. The Moon's disc is almost one-third that of the Earth's. Mighty Ganymede, by comparison, has a diameter less than 4 per cent of Jupiter's. Our Earth–Moon combo is unique in the Solar System, and some believe we should treat it as a binary planet rather than a planet and moon. That said, the centre of mass for the two bodies (the so-called barycentre around which they both orbit) lies within the Earth, which tends to scupper the Moon's claims to planethood.

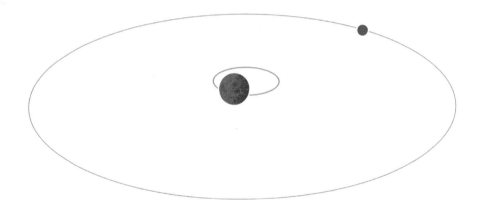

Pluto (centre) and its moon Charon orbit a point in empty space at which their masses balance.

Hold on. Barycentre? Let's back up a bit, as this is an intriguing and important concept that often gets skipped when pondering the planets. It's also quite a cool phrase. The word might be unfamiliar, but we all have an intuitive grasp of barycentres. Try balancing your phone on the end of your index finger. The spot where it holds steady is the centre of mass. A barycentre is much the same. It is the point in space where two or more orbiting bodies would balance if you could somehow hold up the system of spinning orbs with a godlike, multi-dimensional finger. It is also the point around which those bodies are orbiting.

And here's the crucial bit. The barycentre will be closer to the object of greatest mass, but not right at its centre. With the Earth–Moon system, as already noted, the point is within the Earth – about three-quarters of the way out from the centre of the planet. Both the Moon and the Earth orbit this point, but while the Moon sweeps round, the Earth merely wobbles. In other systems, the barycentre is not so neatly buried. Take Pluto. Its main moon of Charon has a comparable mass to its parent planet. The two dance around a mutual barycentre 960 km (597 miles) *above* Pluto's surface. Hence, it would be misleading to say that Charon orbits Pluto – they both orbit a point in empty space, close to Pluto.

Finally, it's wrong to think of planetary orbits as circular, with the Sun in the middle. That one got quashed some 400 years ago thanks to Johannes Kepler. All the planets in the Solar System describe elliptical paths around the Sun. For the inner planets, like Earth, the ellipse is fairly close to a circle. Uranus and Saturn, meanwhile, move around in pronounced ellipses, where the Sun is not at the centre. As noted elsewhere, Pluto's route through the heavens is so eccentric that it passes within the orbit of Neptune for part of its cycle. Even wilder orbits have been observed around other stars.

If all this talk of orbits makes your head spin, at least you'll now be able to phrase it in terms of barycentres.

Now we've sent a probe to Pluto, the whole Solar System has been explored

Distant Pluto* has fascinated us ever since its discovery in 1930. It is too far away for even the Hubble Space Telescope to get a good eyeful. So when the small world was finally visited in 2015 by a spacecraft, everyone was enchanted. The images sent back by New Horizons were astonishing. Far from being a frozen ball of rock, Pluto was revealed as a world of geological activity, with a nitrogen atmosphere and even the possibility of snow.

---

* FOOTNOTE The demotion of Pluto to a dwarf planet caused a surprising amount of anguish for an object that no naked human eye has ever seen. For decades, every schoolchild learnt the names of the nine planets in the Solar System. Then, suddenly, someone changed this to eight, and we all had to go along with it. Trivia fans like me were also disgruntled because the reclassification ruined a long-cherished piece of nitpicking that would otherwise have merited a section in this book. To anyone who declared Pluto the furthest planet from the Sun, we could say 'Aha, not necessarily!'. Pluto cuts inside Neptune for 20 years of its 248-year orbit. Between 1979 and 1999, Neptune was further from the Sun than Pluto. Not a lot of people know that.

Scientists declared that the first phase of Solar System exploration was now over. Machines from Earth have now visited all seven other known planets and most of their moons, plus Pluto, which was regarded as a planet until 2006. Yet here the words 'first phase' are key. We have barely begun our survey of the Solar System, for there's much more to it than planets and moons.

For starters, there may be lumps of rock out there that are larger than Pluto. In 2005, astronomers spotted a world orbiting the Sun at an average distance twice that of the former planet. Since dubbed Eris, the object has a greater mass than Pluto, although it is slightly smaller in volume. It was the discovery of Eris and other 'trans-Neptunian' bodies that led to Pluto's downfall. Eris might otherwise have been cast as the tenth planet, but then there was a strong likelihood that many comparable worlds would be found. Who knows how large the tally might eventually have grown? So it was decided to demote Pluto to the new category of dwarf planet. It is joined by Eris, as well as the asteroid Ceres and two further trans-Neptunian worlds called Makemake and Haumea. Several other bodies could probably qualify as dwarf planets but have not yet been formally categorized. Ceres was also visited by a space probe, known as Dawn, in 2015. The mission puzzled astronomers and armchair pundits alike when photographs revealed a humdrum rocky world punctuated by small patches of super-bright material, now thought to be salt. All utterly unexpected. So two of the five known dwarf planets have been visited, and both confounded the experts. What might lurk on distant Makemake, Haumea and Eris? And who knows how many other dwarf planets might be skulking beyond Pluto? Some astronomers put the estimate as high as 10,000. Probes reach the outer Solar System, on average, once every ten years. At the present rate of exploration, that would mean it would take something like 100,000 years to survey all of these bodies (although you could speed things up by visiting multiple worlds with each probe).

Dwarf planets are just one category of under-explored objects in the Solar System. Astronomers already know of over 1,000 smaller objects in an area around and beyond Pluto known as the Kuiper Belt. Nobody knows how many more might be found, but the best guess is 100,000 with diameters over 100 km (62 miles), plus millions of smaller objects. Other than Pluto, not a single Kuiper Belt object has been visited by our robots, although the New Horizons probe is on course to encounter one on New Year's Day, 2019.

Such distant objects need not be small, either. In early 2016, astronomers announced evidence for a world ten times the size of Earth, orbiting 20 times further out from the Sun than Neptune. Such a leviathan would surely merit the status of planet. At the time of writing, the mysterious world has not been observed directly, but its existence would neatly explain the eccentric orbits of several smaller bodies in the region. There may be other planetary candidates waiting to be discovered.

Throw into the mix more than 5,000 known comets, millions of asteroids, and many other bodies with exotic names like 'centaurs', 'trojans' and the icy planetesimals of the distant 'Oort Cloud' (none of which has ever been imaged), and we have a very long way to go before we can declare comprehensive knowledge of the Solar System. The more we explore, the more oddities we encounter. Take, for example, the awkwardly named comet 29P/ Schwassmann-Wachmann. Every couple of months, the comet undergoes a spectacular increase in brilliance, occasionally brightening 10,000-fold. Nobody knows why.

Even if we were to map every last square metre of the Solar System, there would still be plenty to learn. Until recently, it was thought that life might only exist in the so-called 'Goldilocks Zone'. This small band, encompassing Earth and Mars, is just far enough from the Sun that water can exist in a liquid state, one of the prerequisites for life as we know it. An increasing number of more distant environments are now thought to fit the bill. Jupiter's moon Europa, for example, almost certainly has a liquid water ocean beneath its icy crust. What might be concealed within? Its neighbour Ganymede holds similar potential, as does Saturn's moon Enceladus. Even the most distant objects in the solar system might be members of the extraterrestrial sub-aquatic club. The minor planet Sedna, way beyond the orbit of Neptune, is thought by some to hold a subsurface water ocean.

The exploration of the Solar System has not been completed with the fly-by of Pluto. To employ an oft-heard quote from Winston Churchill, 'This is not the end. It is not even the beginning of the end. But it is, perhaps, the end of the beginning.'

What is a dwarf planet anyway?

There was no formal definition of a planet until 2006. The discovery of Eris, and the probability of other large bodies prompted a rethink. The International Astronomical Union eventually ruled that all bodies in the Solar System (other than the Sun) could be placed into one of the following categories:

Planet: the most familiar category describing a large, spherical mass orbiting the Sun. There are eight planets: Mercury, Venus, Earth, Mars, Jupiter, Saturn, Uranus and Neptune.

Dwarf planet: these are similar to planets in that they must orbit the Sun (rather than another body), and must have collapsed down to a broadly spherical shape. They are typically smaller than planets, but the main distinction is that dwarfs have not 'cleared their neighbourhood' of random rubble. Ceres, for example, is a sphere in orbit around the Sun, but it shares that region with thousands of smaller asteroids. Similarly, Pluto is just one of many objects in the so-called Kuiper Belt. Full-on planets, such as Earth, have sufficient mass to have long ago mopped up any other bodies in their region, other than their well-defined satellites.

Satellite: anything that orbits a body of greater mass (usually a planet or dwarf planet). Satellites may be spherical (like Earth's Moon), or not (like the two moons of Mars).

Small solar system bodies: a broad term to include everything else flying around the Sun. The category includes comets, all the asteroids except Ceres (which is spherical and therefore a dwarf planet), and the various chunks of rock found in the outer solar system.

A miscellany of unscientific laws and theorems

Not everything dubbed a 'law' has a scientific underpinning. For a bit of fun, here are some everyday truisms that are given the lofty status of laws.

Betteridge's law Any headline that ends in a question mark can be answered by the word *no*. Coined by journalist Ian Betteridge as a rejoinder to headlines that ask loaded questions. Common examples include 'Have we detected life on another planet?' and 'Is this photo proof that ghosts exist?'. Sadly, the law can easily be undermined by asking 'Is Betteridge's law true?'.

Clarke's three laws Science fiction writer Arthur C. Clarke would regularly flourish his 'three laws', all of which are non-scientific, yet contain a pearl of wisdom.

- **First law** When a distinguished but elderly scientist states that something is possible, he is almost certainly right. When he states that something is impossible, he is very probably wrong.
- **Second law** The only way of discovering the limits of the possible is to venture a little way past them into the impossible.
- **Third law** Any sufficiently advanced technology is indistinguishable from magic.

Godwin's Law As an online discussion grows longer, the probability of a comparison involving Nazis or Hitler tends towards certainty. Named after American attorney Mike Godwin, the law dates back to the basic newsgroup discussion boards of the early 1990s. Anyone who gets sucked into an online comments thread will find that this law still holds true today. Godwin's Law is now so commonly invoked that it has found its way into the Oxford English Dictionary – anyone who thinks it hasn't earned its place is a language Nazi.

Moore's Law The famous observation that the number of transistors in an integrated circuit doubles every two years. Gordon E. Moore made the prediction in 1965 and it has more or less held true ever since. In part, this is self-fulfilling prophecy – the industry has used Moore's Law as a goal for development and perhaps the pace would not have been so rapid were it not for that initial target. In addition, there is no logical reason why the law should continue to hold much further into the future. We are reaching the physical limits of chip architecture, unless some radical new approach becomes economical.

Murphy's Law Also known as Sod's Law. Anything that can go wrong, will go wrong. It is attributed to aerospace engineer Edward A. Murphy Jr. The most frequently quoted example states that a dropped slice of toast will always land butter-side down – effectively making it inedible. That's not necessarily a bad thing: if you attach the toast butter-side up to the back of a cat, you can build an anti-gravity device and perpetual motion machine. Why? Because cats always land on their feet while, by Murphy's Law, the toast must always land butter-side down. Strap the two together, and the cat-toast entity will spin above the ground as both sides attempt to land first. Whole caverns of the Internet are devoted to the subject, should you wish to explore further. Murphy's Law has also spawned its own parodies. Muphry's Law, for example, requires that 'If you write anything criticizing editing or proofreading, there will be a fault of some kind in what you have written'. Given the title of this book, I hope that law does not extend to factual criticism.

Stigler's Law No scientific discovery is named after its original discoverer. See Wrong inventor?, page 70.

The edge of Physics

The universe is not only stranger than we imagine,
it is stranger than we can imagine.

We live in a four-dimensional universe

Since Einstein, we've all become pretty comfortable with the idea of four dimensions – three of space, plus one of time. Yet it's perfectly possible, even probable, that other dimensions lurk out there, or in there, or around there, or … well, the vocabulary gets a bit tricky. Where does one fit an extra dimension?

Theoretical physicists are no better at observing additional dimensions than the rest of us, but they can use maths, twisted algebra and strange geometries to describe such things. Perhaps the most famous approach is called superstring theory, so named because it posits the existence of tiny vibrating strings, which make up the fabric of reality. The real driver of superstring theory is to understand how gravity works at the most fundamental of levels, and to mathematically wrangle the stubborn beast into an equation with the other three forces of nature. Gravity will not be corralled into any old pen, and physicists must conjure additional dimensions to squeeze it in. One of the leading versions of string theory requires ten to explain creation. Yes, ten.

It's not easy to get one's head around these ten dimensions, but I'm up for it if you are. The following description takes a few liberties with the vocabulary – everyday English is not really adequate for such high concepts – but should give you some notion of what we're dealing with.

The first three dimensions are easy: length, breadth and depth. We also have an intuitive (if flawed) sense of the fourth dimension, time. For what follows, imagine time in a broad sense, as a straight, one-dimensional line connecting the Big Bang at the start of the Universe to whatever awaits us at the end of the Universe. Got it? Good. We're now going to venture into the higher dimensions.

To enter the fifth and sixth dimensions, we need to elbow time away from its one-dimensional progress into other directions. It's not as hard or violent as it sounds. Every time we make a decision, we pass along a distinct timeline. A journey through the fifth dimension would allow us to move between two alternative timelines. Imagine a scenario where you're listening to a particularly difficult talk about physics. The lecturer is making no sense, and is really quite irritating. You remember that you have a water pistol in your pocket. Do you give the professor a soaking, or keep your cool and let him waffle on? As a four-dimensional being, you can only do one or the other, and face the consequences. A five-dimensional audience member would have the giddy pleasure of spraying the lecturer, then travelling to the alternative timeline where no such dousing occurred.

It's very much like adding a second dimension to time. The sixth dimension, by analogy, can be likened to three-dimensional time. It includes any and all possibilities that might occur in our Universe. Instead of soaking the speaker, you might have shouted for him to stop, or walked out, or uploaded a #BoringProf photo to Instagram. The sixth dimension is the space that includes all these possibilities plus everything else that might happen throughout the Universe at any point in its existence.

That's still only six dimensions. How can you possibly have even higher regions that go beyond all possible outcomes? So far, our inter-dimensional wandering has been limited to regions where the rules make sense. When you squirted the physics professor, your jet of water didn't float up to the ceiling or speed down to the ground. Nor did your target disperse in a cloud of atoms. The first six dimensions don't take us anywhere too exotic. Stroll where you will, the fundamental forces of nature will remain the same.

To enter the seventh dimension, we must step outside all of that. Picture a universe that begins with a Big Bang, just like ours, but in which the forces of nature are slightly (or hugely) different. For example, gravity might be 8 million times as attractive, or one-hundredth as weak. The so-called strong interaction, the cement that holds the building blocks of atoms together, might reach out further in our destination universe, creating particles utterly unlike anything we know here. In short, a trip to the seventh dimension takes us to a space where the four fundamental forces of nature could have wildly different values.

To take things higher still, imagine a plane
that includes every possible universe, with
every permutation of the four forces.
That is the eighth dimension.

Queasy travellers may
wish to leave now,
because things are
about to get even
more warped. The
ninth dimension brings
together the idea of all the
possible starting points, with
all the possible futures. It allows
us to move between any physically
possible state or outcome in
this Universe and any other in
another universe. The tenth, and
highest, dimension is the complete
repertoire of all these states
combined together.

Phew.

Nobody, of course, has ever
measured, witnessed or travelled to
any dimension higher than the fourth.
These are all theoretical dimensions.
Yet all ten are necessary to make string
theory work. If we wanted to jettison the
ninth and tenth dimensions because, well,
they're ludicrous, then the maths simply
wouldn't hold up. The situation would be just
about acceptable, were it not for the fact that
there are alternative versions of string theory. Some

need 11 dimensions, one requires as many as 26. I don't think I have a hope of describing all those, so I'll instead refer you to an alternate version of this book, accessed via the fifth dimension.

Nothing can travel faster than light

This is one of those facts of science that everybody can repeat. Since Einstein, we've known that nothing can travel faster than the speed of light. It's not just that light can move pretty darn fast – roughly 300,000 km (186,000 miles) every second – but also that, fundamentally, the laws of physics do not allow anything to surpass that speed. A spaceship would need an infinite, and therefore impossible, amount of energy. That said, there are a few cheats we can employ to get around this limit.

Under the right conditions, a wheezing tortoise can outpace a beam of light. It all depends on the medium. If we're talking about deep space, where there's little to get in the way, light will zip along faster than you can possibly imagine. This is true not just of the light we can see with our eyes, but also X-rays, gamma rays, radio waves and other forms of radiation. But stick something transparent in front of the light, and we can clip its heels.

The Earth's atmosphere, for example, taps the brakes a little. Passing through the layers of air slows the Sun's rays by some 90,000 m/s (27,000 ft/s). That sounds like quite a wrench, but it's only 0.03 per cent slower than the speed in a vacuum. Glass and water cause a much greater retardation – slowing the light by around one-third. This change in speed also causes a change in angle called refraction – an effect readily experienced from the bottom of a swimming pool.

If you can get your hands on much more exotic materials than glass and water, then you can slow light down still further. Researchers hit the headlines in 1999 when they hobbled light down to just 17 m/s (56 ft/s), or 61 km/h (38 mph) – your car shifts faster than this on a dual carriageway. The researchers used a cloud of supercooled sodium atoms known as a Bose-Einstein condensate.

More recent advances have been able to bring light to a standstill, and store it for future release (unlike when light hits a wall, and is simply absorbed or reflected). Such light traps have enormous potential for communications devices and storage of data.

In the past few years, scientists have even tricked light into slowing down in a vacuum. The technical details are complex, but in simple terms, particles of light known as photons are sent through a mask that alters their 'shape', slowing them down ever so slightly. The particles retain this diminished speed even after leaving the mask. Effectively, particles of light in a vacuum are travelling at less than the speed of light in a vacuum, which was supposed to be a constant.

There are other cheats for getting around the light barrier. One, the warp drive, is familiar from science fiction. Rather than accelerating beyond the speed of light, the USS *Enterprise* bends space to bring the destination closer. Nothing in Einstein's theories of relativity prohibits this approach – we just don't know how to do it yet. Similarly, the idea of using a 'wormhole' to tunnel between two distant regions of space – and thereby make the journey between them faster than light – is also a theoretical possibility. No wormhole has yet been observed and nobody knows how to create one.

One might also look to the quantum level for an example of an apparently faster-than-light process. It's possible to take two particles and buzz them into the same quantum state so that they are said to be 'entangled'. Everyday words are inadequate to describe exactly what this means, but essentially, the two particles then share one existence. If you then send the twins off in opposite directions, they will remain entangled whatever the distance. Perturb one of the pair, and its partner will instantaneously be affected. This strange result, which Einstein described as 'spooky action at a distance', appears to break the light speed barrier. However, the set-up is so imbued with randomness that it is impossible to use it for any form of communication, or to pass on any data.

A final and entirely frivolous exception to Einstein's rules is the British Royal Family. The royal line has a clear rule of succession. As soon as a monarch dies, the next person in line to the throne will immediately become king or queen – hence the chant 'The King is dead. Long live the King'. The transfer of title is instantaneous, and can therefore be considered faster than light. If the future

William V happens to die on Mars, his son Prince George will immediately be the rightful King – although he won't know it for the 20 minutes or so it takes the sad news to travel, at the speed of light, back to Earth from the red planet. The notion is described in law as *le mort saisit le vif* ('the dead seizes the living'), and may be applied to other cases of inheritance.

Nothing can escape a black hole, not even light - so they can't be detected

Hopefully, you've been reading for a while now, and are due a break. If you're able, why not stand up and stretch your legs for a few minutes, then come back. Done? Good work. You just raised your mass a few centimetres further from the centre of the Earth. For all the bulk of the planet, its gravitational attraction isn't all that impressive. Most people's leg muscles are strong enough to counter the Earth's pull, at least over a short distance. You might like to jump in the air and emit a loud 'Whoop!' in celebration.

Try the same trick on a larger planet and you wouldn't be so cocky. If we could double the Earth's size, it would be eight times as massive, and would pull at our bodies with twice the force of gravity. You'd probably still be able to get out of your chair, but you might not be so inclined to whoop. Make the planet bigger and you'd never raise from your chair. Larger and the chair would collapse, along with your skull and ribcage. Take things to the extreme, and all matter on the surface of the planet would be squeezed inward. No driving force could push your remains up from the irresistible centre. You have entered a black hole.

Black holes are absolute monsters when it comes to gravity. To reverse the last thought experiment, imagine that you could take the Earth and crush it down so that it would fit in your coffee mug, yet still keep the same mass and gravitational attraction. That's the kind of hyperdensity that black holes enjoy.

Your coffee mug would quickly be consumed, along with your hand, arm, torso and anything else within sight.

Almost any book or article that describes a black hole will use precisely the same phrase to describe its effects: 'nothing can escape, not even light'. The notion is, as far as we know, entirely correct – once you are beyond the event horizon of the black hole, there is no coming back. That said, there's an interesting twist in the physics that allows these objects to emit radiation. In other words, black holes are not black, and radiation can, in some sense, escape.

It all has to do with a phenomenon known as virtual particles. Quantum mechanics shows us that so-called empty space is not empty at all. It seethes with virtual particles – that is, particle-antiparticle pairs that pop into existence and then immediately annihilate one another. It happens all the time.

In 1974, Stephen Hawking described what would happen to the virtual particles that flit into existence just outside the event horizon of a black hole. One particle would fall into the event horizon of the black hole, never to be heard of again. The other, suddenly lacking its partner for self-annihilation, would remain in the Universe, and might even radiate away from the black hole. Collectively, these escapees are termed Hawking radiation. It is an effect that has yet to be observed, but carries strong credibility among theoretical physicists.

Now, Hawking radiation isn't truly breaking the maxim that nothing can escape a black hole. It originates from just outside the event horizon, not quite the point of no return. Nevertheless, its probable existence overturns notions that black holes are entirely black and therefore do not emit radiation. We can only speculate – with maths and physics – what goes on beyond the event horizon. Here, the laws of physics break down.

At the time of writing, nobody has detected Hawking radiation, nor taken an image of a black hole. We have almost incontrovertible evidence for their existence, including the discovery, in early 2016, of gravitational waves thought to come from the catastrophic embrace of two black holes. Most commonly, though, the game is given away by the swirling clouds of gas, attracted to the huge gravity sink but not yet swallowed. These emit distinctive X-rays, which have been detected from Earth.

Wrong inventor?

Invention is the appliance of science, when trial, error, experimentation and inspiration lead to a product that can be used by others. For some reason, humans love to hear tales of lone geniuses and maverick tinkerers who single-handedly change the world with their ideas. In reality, science and invention rarely work like that. The people we habitually credit with an invention have often only played a part in its genesis. There are three main reasons why we might make this mistake.

- Many inventions are the culmination of thousands of hours of research and design, often by large teams of collaborators. There is no single inventor, only a team leader or figurehead who gets most of the press. Think, 'Steve Jobs invented the iPhone'.
- Inventions seldom come out of the blue. They are usually improvements or twists on an existing technology. A chain of inventors each contributes some improvement to a device; any one of them might be declared the inventor.
- Quite often, an inventor is unable or reluctant to take their creation to market. A few years later, somebody else comes along, takes up the same idea (either independently or deliberately), and gains fame and fortune from their successful enterprise. That second person is often credited as the inventor.

The whole tangled mess is summed up in the adage known as Stigler's Law: 'No scientific discovery (or invention) is named after its original discoverer'*. Here are a few of the more famous and clear-cut examples from history.

Thomas Crapper invented the flushing toilet. Everybody wants this to be true because ... well, *that* surname. It's not true, though. The water closet is usually attributed to John Harrington (hence, 'going to the John'), some time in the 1590s, and the ensuing centuries are flush with variations on his design. Crapper, working in the late 19th century, was more of a salesman, peddling bathroom furniture from his London showroom, though he did hold several toilet-related patents. The word 'crap' as a term for waste or defecation also predates Mr Crapper, though his ubiquitous branded toilets must have helped cement the slang.

Thomas Edison invented the light bulb. One of the world's most disputed stories of invention concerns the humble light bulb. Its genesis is often attributed to Thomas Edison but, as so often with new technology, he was just one player in a crowded workshop. Alessandro Volta might get some credit, as the first to make copper wire glow due to an electric current, in 1800. Two years later, and Humphrey Davy had a working 'arc lamp', a bright light formed when ionized gas was placed between carbon electrodes. A more recognisable light bulb was invented in 1840 by Warren de la Rue. His choice of platinum for the element was prohibitively expensive – hence why he's not a household name. Various other competing designs surfaced over the following decades. The most notable came from Joseph Swan, who had a working bulb by 1860, and a British patent on an improved design by 1880. Edison bagged the US patent on a similar design and successfully took it to market. Any one of the names above (and others besides) might be championed as 'the inventor of the light bulb', and whole books have been written in an attempt to shed light on proceedings.

---

* FOOTNOTE Thomas Stigler, after whom the law is named, attributes the phrase to sociologist Robert K. Merton, thus proving that Stigler's Law follows Stigler's Law.

Henry Ford invented the motor car and assembly line. Automobiles had been around as curiosities long before Ford introduced the Model T in 1908. Horseless carriages powered by steam date back to the 18th century, while the first autos powered by internal combustion engines hit the roads in the 1880s. Rather, Ford was the first to make a successful business out of selling vehicles to the middle classes. He did this through efficient assembly lines. The growing automobile would move along a conveyor for the attentions of a series of workers, each with a specialization. It's often claimed that Ford invented this methodology, but conveyor-belt manufacture had been around for many years by the time he adopted it. Notably, another car manufacturer, Ransom Olds, had an efficient assembly line several years before Ford.

Al Gore invented the Internet. The former Vice-President of the USA was widely lampooned after he claimed to have invented the Internet. The inconvenient truth is that he never made such a boast. In a 1999 interview, Gore said that he '... took the initiative in creating the Internet' while serving as a congressman. His meaning, clear from the wider context, was that his political support had been a key factor in the growth of the Internet. He was not claiming to have invented the technology. It's like saying 'President Kennedy took the initiative in creating the US Moon programme'. Nobody would take that to mean that Kennedy invented the rockets or worked out the orbital mechanics that got men to the Moon. The extent of Gore's role in fostering the Internet might be debated, but he certainly didn't invent the thing, nor claim to.

Pythagoras spotted a relationship between the sides of a triangle. We all learn the Pythagoras theorem at school: for any right-angled triangle, the sum of the squares of the two smaller sides equals the square of the longest side. This equation was known centuries before Pythagoras, and was utilized by many civilizations. Pythagoras did not discover it, but was (probably) the first to prove that it holds for all cases.

James Watt invented the steam engine. Watt is rightly held in high regard for his singular contributions to the steam age, but he wasn't the first to develop a useful engine. The little-known figure of Thomas Savery is perhaps best deserving of the accolade. In 1698, he received a patent for a device capable of 'raising of water and occasioning motion to all sorts of mill work'. In reality, the engine was not very powerful and found only limited use. The first commercially successful steam engine was constructed by Thomas Newcomen in 1712, and built on the designs of Savery and others. The Newcomen engine powered the early industrial revolution for around 70 years before James Watt came along with a much more economical solution.

Curious Chemistry

Adventures in the world of atoms and molecules,
where nothing is quite as you'd expect.

CH₂ ⁺
O

Chemicals are bad for you, and should be avoided. Eat naturally!

Do not, whatever you do, ingest dihydrogen monoxide. This colourless, odourless substance is a prime component of acid rain, contributes to erosion of the natural landscape, causes metals to rust and is a leading solvent for industrial processes. Thousands of people die each year after inhaling this substance. And yet manufacturers continue to add it to most consumer products.

Dihydrogen monoxide is, of course, more familiar to us as H_2O or water. Scaremongering around dihydrogen monoxide is a well-known hoax, one designed to show how language can be twisted to play on people's fears: in this case, that natural = good and chemicals = bad. We readily accept that water can do all the things claimed above, but dress it up as dihydrogen monoxide and it suddenly looks evil.

Society at large is very wary of chemicals. This is a healthy stance to take. After all, if chemicals like DDT and CFCs can screw up the environment, what might others do to our bodies? But we shouldn't assume that all chemicals are bad, just because a few are. It is a logical fallacy – like saying that sharks sometimes kill people so we should avoid all fish. Besides, everything we consume is made up of chemicals. Not just water, but fat, protein, sugars, carbs and vitamins. Breakfast, lunch, dinner and supper: all of it is chemicals. Hence, to avoid eating chemicals is to go on the most limiting form of diet possible.

I'm being deliberately facetious, of course; those who tell us to avoid chemicals are really talking about additives and E-numbers. While a handful of common additives should be avoided by those with certain medical conditions, the vast majority are harmless (or even beneficial) in everyday amounts.

Perhaps the most maligned additive is monosodium glutamate (MSG). This white powder is used to enhance savoury flavours, and is particularly associated with Chinese cuisine. It has been linked to any number of woes including headaches, numbness and feelings of hunger not long after eating. Even the name monosodium glutamate looks synthetic and evil.

In reality, MSG is a natural and abundant compound, found in many fruits, vegetables and dairy products. As such, humans have always eaten it, and we've been adding extra quantities to our dishes for over 100 years. Despite plenty of anecdotes, no scientific study (and there have been dozens) has ever found a conclusive link between MSG and any of the reported side effects. Sure, if you eat a ludicrously large dose, you're going to feel queasy. But time and again, research has found that MSG is perfectly safe to eat in reasonable quantities.

Another key example is aspartame, an artificial sweetener commonly used, for example, in diet soft drinks. The Internet is puddled with misinformation about aspartame. It has been linked to everything from cancer to Alzheimer's disease to birth defects. Many websites repeat the warning (word for word, which should always raise a red flag) that aspartame is 'by far the most dangerous substance on the market that is added to foods'. Such alarmist messages are completely at odds with the weight of scientific evidence. The US Food and Drug Administration reckons aspartame to be 'one of the most thoroughly tested and studied food additives the agency has ever approved', and its safety is 'clear-cut'. Aspartame's problems apparently started in an Internet hoax, which was widely believed and repeated. The only people who really need to dodge the chemical are those with phenylketonuria, who must avoid high levels of phenylalanine, one of the constituents of the sweetener. For everyone else, aspartame is safe, and a healthier alternative to sugar.

MSG and aspartame are just two of many examples from what might be called the 'don't touch it, it's not natural' fallacy. The argument that we should avoid something because it's 'not natural' is quadruply problematic.

- Many man-made substances are stupendously useful and beneficial. Nothing in nature can compete with the products we use to brush our teeth, wash our hair, treat cancer or block headaches.

- At the same time, many compounds we might assume are man-made are, in fact, natural. Some of the most common preservatives – sorbic acid, benzoic acid, sulfur dioxide, natamycin and various nitrates – all sound viciously man-made, but occur abundantly in nature.

- It's true that some synthetic compounds are best avoided. Equally, many natural or organic compounds are also harmful. Salt, for example, is one of the most abundant compounds on the planet. It figures prominently in most people's store cupboards. Yet its over-consumption contributes to millions of deaths a year from various heart-related diseases. Most poisons and narcotics are also natural in origin.

- What do we mean by 'natural' anyway? Is organic food natural*? Most organic crops are grown in large fields, sometimes in monoculture, usually with fertilizer, to be harvested by tractor. Is that natural? Is wearing clothes natural? Driving a car is surely not natural – or is it just advanced tool use? Those who advocate natural living often do so by typing away on a plastic keyboard beneath a panel of liquid crystals, before uploading their thoughts into a network of cables, transistors and electromagnetic pulses – just, one must presume, as Mother Nature intended.

---

* FOOTNOTE There may be good environmental and ethical reasons for eating organic food, but the human health benefits are not so clear-cut. Many studies have compared organic and non-organic food, with mixed results. Some show a minor nutritional benefit for organic, while others do not.

The natural versus synthetic debate is finely nuanced. It is folly to avoid all chemicals and embrace only those found in nature. Each and every substance we consume should be considered on its own merits, and not according to its classification as man-made or natural. Just because something is natural does not make it desirable or safe. Just because something is non-natural, does not make it inherently dangerous.

Water boils at 100°c and freezes at 0°c

Measuring temperature in Celsius (°C) rather than Fahrenheit (°F) makes sense for two reasons. First, Celsius is much easier to spell. Second, the scale maps so neatly onto the freezing and boiling points of water. As we all know, water freezes at 0°C (32°F) and boils at 100°C (212°F).

Or does it? Well, yes, but only under very, very specific conditions. Boiling and freezing points are fickle things. They change with altitude or, more accurately, pressure. If you were lucky enough to own a penthouse in the Burj Khalifa (the world's tallest building, at around 828 m/2,717 ft), yet still humble enough to make your own pasta, you'd find your pan of water boiling around 97°C. Move your kitchen to the top of Everest, and you can achieve a gentle simmer at a mere 70°C.

While you're enjoying your pie in the sky, it's worth noting that the effect of altitude on boiling point is a real factor for many people. The 50,000 citizens of La Rinconada in Peru, for example, perch some 5,100 m (16,728 ft) above sea level. The water in their kettles boils at 83°C. The two million inhabitants of La Paz, the administrative capital city of Bolivia, must settle for tea at 88°C. Such differences matter when preparing food. Charles Darwin grumbled about his undercooked potatoes while travelling through the high regions of Argentina. Modern inhabitants of lofty cities use pressure cookers to mitigate the effects of altitude.

Even if you live down at sea level, your water will probably never boil at 100°C. By international definition, your appliance must feel the mean atmospheric pressure at mean sea level at the latitude of Paris, which turns out to be 101,325 Pascals in the official units of pressure. It is a rare kettle indeed that precisely matches this pressure. If you live at or around sea level, then you probably get pretty close to this most of the time, but very rarely, if ever, will your water boil at precisely 100°C.

And let's not forget the rest of the Universe here. So far, we've discussed only the surface of the Earth. In terms of the wider cosmos, this is a vanishingly non-representative environment. Almost all of existence, from low-Earth orbit to the edges of the Universe, hovers around -270°C and in near vacuum. What happens to liquid water here? It should freeze at such a low temperature. On the other hand, the pressure is practically non-existent, so the boiling point is greatly reduced. Does the water freeze or boil? It kind of does both. An astronaut peeing into the void would see their ejected stream evaporate into a gas, and then instantly 'desublimate' into golden crystals. All rather beautiful.

Pressure isn't the only thing to swerve the boiling point of water. Purity is also a factor. Tap water is full of dissolved minerals and other impurities. These have the effect of raising the boiling point ever so slightly above 100°C. Throw some salt into your pan of spaghetti and you can get the temperature up a degree or two above the standard boiling point. Freezing point, meanwhile, is lowered by impurities. This is why your local authority spreads salt over the roads in cold weather. With salt present, the temperature needs to drop lower than 0°C before ice will form on the road (the grit also gives a rougher surface for tyres to grip).

So when we say that water boils at 100°C, or that it freezes at 0°C, we are talking about ideal conditions. In the mess of reality, it rarely does.

Matter exists in one of three states: solid, liquid or gas

Most things in everyday life are easily classified as a solid, liquid or gas. We can intuitively tell them apart. Nobody but a fool would try swimming through a cloud of steam or flushing their toilet with ice.

A solid object contains atoms or molecules that are packed closely into regular, repeating structures. Supply a bit of heat and the bonds holding this arrangement together begin to vibrate and then break. At our human scale, the substance seems to get softer and melt, as when ice turns to liquid water. At the minute level, the strong bonds are broken, and the molecules are free to move around. Molecules in a liquid still feel a tug from neighbouring molecules, but the attraction is much weaker than in the solid. If the liquid is heated further, some molecules obtain enough energy to break away from the mass. They evaporate, or turn into a gas. Molecules in a gas have lots of freedom to move around and feel little attraction to neighbouring molecules – hence you can walk through a gas, wade through a liquid but cannot pass through a solid without breaking a few ribs.

We tend to hang around in places with such agreeable temperatures and pressures that we never notice anything other than solids, liquids and gases. But there is a fourth fundamental state of matter, and one that is more abundant than the other three put together. It's called plasma.

Plasma is what you get when you rough up a gas, usually by heating or buzzing it with electricity. This rips electrons away from their nuclei, creating a tenuous soup of charged particles. For this reason, plasmas are sometimes known as ionized gases (an ion being another name for a charged particle). Plasmas share physical similarities with gases – they have no fixed shape, for example. The mix of charged particles gives a plasma very different properties in other respects. They readily conduct electricity and respond to magnetic fields. They can also form into filaments.

Sounds exotic, right? Well, sort of. But plasmas are much more familiar than you might think. You may well have spent hours staring at the stuff if you own a plasma TV. The steady glow from a neon lamp is also produced by plasma, as is the sharp flash of lightning. If you've ever played with one of those weird glass balls in which coloured tentacles of energy emanate from a central bulb to your hands, then you've almost touched plasma.

The plasma we encounter more than any other, however, is the Sun. Much of the matter within our star – and all stars – is in the plasma state, with hydrogen stripped back to its constituent protons and electrons. Even greater quantities of plasma exist in the great 'voids' between galaxies. Here, vast filaments of ionized hydrogen extend for thousands of light years through the heavens.

So, we have four states of matter that you or I can observe any day of the week: solids, liquids, gases and plasmas. The phases of matter do not stop there, however. Matter can exist in many other states – it's just that we don't often encounter them. A Bose-Einstein condensate, for example, is a gas that's been cooled to just a few billionths of a degree above absolute zero – much colder than the typical temperatures of deep space. At such temperatures, the individual particles become so sedentary that they lose all sense of individual identity and behave as one super-particle. You won't find a Bose-Einstein condensate down the back of your sofa, unless you have peculiar living arrangements. You need a Nobel-Prizewinning laser rig and a team of capable physicists to observe one. Nevertheless, it's a perfectly valid state of matter, and one that can teach us much about the fundamental laws of the Universe.

The menagerie of exotic states grows with each decade. Quark-gluon plasmas, superliquids, supersolids, photonic matter and strange matter (a technical term, rather than a lazy description). All of the preceding discussion has ignored the mysterious 'dark matter', thought to make up 83 per cent of the mass of the Universe. Almost nothing is known about dark matter. We can't see it or measure its properties, only infer its existence from its gravitational effect on other matter.

If you'd always assumed that the Universe was made up of solids, liquids and gases then you could hardly be more wrong. Something like 99 per cent of the observable Universe is plasma, with unfathomable quantities of dark matter yet to be observed. To a close approximation, solids, liquids and gases don't exist.

Water is a good conductor of electricity

'A 16-year-old Sheffield boy has met his death in a remarkable fashion. He was having a bath at the same time as two other lads were fixing a blind in the bathroom. They knocked down an electric lamp, which fell with the wire attached into the bath. The lad got so severe an electric shock that he died almost immediately. The current was one of 200 volts.'
Beverley and East Riding Recorder, Saturday 19 February 1916

Tragedies like the one above highlight the dangers of mixing water and electricity. It's a life rule we're all familiar with. But contrary to experience, water is a truly terrible conductor of electricity. Pure water, that is. The wet stuff we get from our taps or find at the seaside is actually a soup of impurities, and it is these impurities that pass on the electric charge. Remove them, and the pure water is an insulator rather than a conductor. To see why, we need to do a brief refresher on atomic structure.

To form an electric current, you need to encourage a flow of charged particles. Copper wire is an excellent accomplice for this. Each copper atom within the wire is surrounded by 29 electrons; 28 of these are strongly bound to the copper nucleus – they're part of three tight-knit gangs, who don't like leaving their home territory. The twenty-ninth electron finds itself alone on the edge of the atom. It feels less of a tug from the nucleus and will readily leave home for another copper atom. In this way, the copper wire has many 'free' electrons randomly and promiscuously circulating among its constituent atoms. If a voltage is applied across the wire, then the free electrons stop moving randomly and all go in the same direction – because they're negatively charged, they find themselves repelled by the negative electrode and head towards the positive electrode. The wire is now said to be conducting electricity.

Electricity isn't only conveyed by electrons. Remove electrons from atoms and the atoms become positively charged. These so-called ions can then be used to conduct electricity – in this case the current will flow towards the negative electrode.

Pure water contains very few charged particles. Occasionally, a water molecule will lose a hydrogen ion, which then hooks up with another water molecule to produce H_3O+ (leaving behind some $OH-$). So there are some charged particles in water, but not many, and not enough to make it a good conductor. All that changes if you add a few salts. Tap water contains plenty of dissolved magnesium and calcium, which hang around in the water in a charged state. These are present in significant enough quantities to propagate the electric current.

You could quite happily share a bathtub with an electric lamp, so long as the water remained super-pure*. With tap water, you'd get a nasty, and perhaps fatal, shock. Seawater, which contains lots of sodium and chloride ions from all that salt, is an even better conductor, as the great white shark in *Jaws 2* found to its cost.

---

* FOOTNOTE DO NOT TRY THIS AT HOME.

Glass is a liquid

Walk around an old building with a know-it-all friend, and they're sure to refer you to the ancient windows. 'See how the glass is thicker at the bottom?', they'll beam. 'That's because glass is a liquid, not a solid. This window has been here so long that gravity has caused the glass to flow downwards'. Glass might appear solid, they'll maintain, but it's really a very, very thick liquid.

As urban myths go, this one appears to be shatterproof. It has long been discredited but is still widely believed. Glass* is not a liquid at normal temperatures, and it does not flow to any discernible degree, even over centuries. Some ancient panes really are thicker at the bottom, but that's because they were made that way.

Medieval glassmakers would boggle at today's glass office blocks. Those of us who live in big cities stroll past such edifices with indifference, perhaps even irritation at how common they've become. To the 13th-century glass blower however, these crystalline cliffs would be as alien and unfathomable as a passing helicopter. The technology of the time could only produce small panes, and these were of variable thickness and transparency. It made sense to install these imperfect panels so that the heavier edges were at the bottom. This, then, is the reason that old windows appear thicker at the base.

There is, however, some support for the notion of flowing glass. If you visit the University of Queensland in Australia, you can witness the world's longest-

---

* FOOTNOTE Just to be clear, we're talking about the silica-based glass commonly used in windows here. Many other forms of glass are known, some with liquid-like properties.

running scientific investigation. The Pitch Drop Experiment was set up in 1927 by Professor Thomas Parnell, to show students that apparent solids can behave as liquids. His sample of tar pitch (or bitumen) is so thick that it only drips once a decade. It is estimated to have a viscosity 230 billion times that of water. If you want an expression for boredom that would trump 'watching paint dry', you might try 'waiting for pitch to drip'.

If pitch can flow so ponderously, then why not glass? At the structural level glass resembles a liquid; its atoms are strewn haphazardly, like toys on a nursery floor. But these atoms are so strongly bonded together that they can't slip and slide past each other. The result is a rigid object known as an amorphous solid. At reasonable, human temperatures, there's no way that such a strong material is going to flow unless you're prepared to wait for millions of centuries. Pitch, on the other hand, is a complex mix of large molecules. It is a soup so gloopy as to appear solid, but given enough time (decades), the individual components can slip past one another betraying its liquid state.

Atoms are like tiny solar systems, with electrons orbiting nuclei

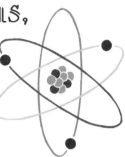

Every atom in the Universe is built on the same principle. Each has a central blob called a nucleus. Small balls called electrons whizz around this nucleus in tightly defined orbits. It's very tempting to liken this set-up to a miniature solar system. The dense, heavy nucleus is like the Sun, while the attendant electrons resemble planets.

You were probably taught something similar at school. The analogy with the Sun and planets is a useful bit of hand-holding when we first learn about atomic structure, yet this picture is ultimately misleading. Atoms are unlike any objects we encounter in our day-to-day lives. The rules are different down at this tiny scale; they are quantum. Particles can pop in and out of existence, or simultaneously exist in two alternative states. It's small wonder that many teachers ease their students into this world with comfortable analogies of solar systems and orbiting spheres.

My chemistry teacher* had a different approach. He avoided the solar system cliché with a macabre analogy more likely to excite the skittish minds of teenage students. Imagine, he'd say, that you squashed a bluebottle fly beneath your thumb, and smeared its body across the table surface. Where are its eyes? Without further information, each eye has an equal probability of being anywhere within the smear. Your crushed fly is, he'd continue, much like a helium atom. Helium has two electrons. They're somewhere in the vicinity of the atom's nucleus, but we can't say for sure exactly where. In effect, they're smeared out into a sphere of probability.

Such spheres of probability are known to chemists as s-orbitals. That word looks a lot like 'orbit', which no doubt has helped reinforce the idea that the electron is going round and round the nucleus like a planet. Any such notions are quickly dispelled, however, when we look at atoms with more than two electrons (anything heavier than helium). Here, the electrons aren't confined to a sphere, but inhabit oddly shaped regions of space, like dumbbells or flower petals. These are known as p-, d- and f-orbitals. Whoever saw a solar system that looked like this?

Rough sketch of three types of f-orbital. The nucleus, not pictured, would be at the centre of each. They look nothing like planetary orbits.

* FOOTNOTE A wonderful chap called Don Ainley, sadly no longer with us.

An atom of plutonium, for example, harbours 94 electrons, occupying dozens of these spheres, petals and lobes. The electrons do not loop around in circular orbits, but are better imagined as smeared-out clouds of probability. Even this is a crude simplification. To more fully understand how electrons behave within an atom, we have to invoke rather complicated mathematics, way beyond the scope of this book. If you're interested, look up the Schrödinger equation.

Life on Earth

Nobody knows quite how life got started, and much
remains mysterious about how it evolved and spread.
Such uncertainty has led to plenty of myths and misconceptions.

Life has only existed for a tiny fraction of Earth's existence

TV science presenter and particle physicist Brian Cox is often lampooned for his frequent use of the phrase 'millions and billions' when describing the wonders of the cosmos. Such teasing is a little unfair – it's hard to avoid incredible numbers when describing the vastness of time and space. The same is true of our planet. Our best evidence, and it's pretty good, puts the Earth at 4.54 billion years old. In other words, a heck of a long time.

Now, we've all seen programmes or read books where our species is shown to be a Johnny-come-lately in the Earth's history. Nobody can intuitively grasp the vast stretches of time – the millions and billions of years – that make up the geological history of the Earth, so more familiar metaphors are often employed. Most commonly, the Earth's backstory is represented as a 24-hour clock, with the formation of the planet on the stroke of midnight. Using this device, mammals appeared on the scene very late in the day, around 11.39pm. Early humans show up with only a minute left on the clock, while us modern humans have known only the last second or two.

It would be wrong to conclude that our miserable insignificance to geological time is true of life in general. We now have fossil evidence of complex, single-celled organisms dating back 1.5 *billion* years – that would be around 4pm on our clock of aeons. Even these creatures are thought to be relative newcomers. The chemical fingerprints of life can be found in rocks of much greater vintage.

Recent evidence may push things back as far as 4.1 billion years ago, when Earth was still a volcanic hell*. A 2015 study measured the quantities of heavy and light carbon atoms in a small crystal from this period. The result gave a ratio that would smack of life, were it found in more recent samples. Further research is needed to definitively pin this one but, if true, it would mean that the planet has been inhabited for nine-tenths of its existence. This 'quick start' for life on Earth offers hope that we might find life elsewhere in the Universe.

* FOOTNOTE This geological aeon, when the Earth was still settling down after its formation, is known as the Hadean, after the Ancient Greek underworld of Hades.

All life depends on the Sun

'The Sun is God,' declared J.M.W. Turner with his dying breath. In some sense, we might all agree with the famous artist. The star at the centre of our Solar System is key to all life on Earth – or so it was thought until recently.

For many years, it was assumed that all living things get their energy from the Sun. Plants and certain micro-organisms benefit directly by exposing themselves to the Sun's rays. Photosynthesis converts the star's radiance into little packets of chemical energy within the plant. Animals are not blessed with the molecular equipment to pull off such a trick. We let the plants do the hard work, and then we eat them. Lettuce-shy carnivores are one step removed, but they still eat animals that were themselves nourished on plant matter. Even species that inhabit the perpetual darkness of the deeper ocean are part of a Sun-begun food chain. These abyssal creatures rely on food sources that have fallen from regions above – poetically known as marine snow – and so ultimately derive their vim from the Sun.

And then, in 1977, a most staggering discovery was made. The deep-sea submersible Alvin* was snooping around an ocean ridge near the Galapagos Islands. The crew were looking for hydrothermal vents, which had been postulated but never observed. Hydrothermal vents are formed when seawater seeps down through the rock and encounters volcanic magma. This kiss of Vulcan superheats the water, forcing it back up through the rock under high pressure. It eventually erupts like an underwater geyser. The divers on Alvin found not only such vents, but also towering columns known as black smokers. These obsidian chimneys grow when water hot enough to melt lead (up to 340ºC) hits the near-freezing temperatures of the abyss. Mineral particles such as sulfites crystallize out, eventually forming tall columns like smokestacks.

The biggest surprise was that these black smokers were teeming with life, much of it previously unknown to science. The divers observed tubeworms, crabs, snails, shrimps, fish and even octopi. All had made their homes in the most crushing circumstances imaginable. From a human point of view, this is an aquatic version of hell: enormous pressures; highly acidic waters; a surfeit of chemicals that most organisms would find toxic; a pitch black abyss where the temperature can vary by hundreds of degrees in just a few centimetres. And yet life thrives here, with far more diversity than the surrounding sea bed.

Bacteria underpin this ecosystem, and form in mats around the vents. They derive all of their energy from the chemicals of the smokestack, and are particularly fond of hydrogen sulfide (you might know it as the smell of rotten eggs). The process is known as chemosynthesis, an alternative energy pathway to photosynthesis. If the Sun blinked out tomorrow, these tiny life forms would carry on thriving. Larger animals at the vents consume the bacterial mats in much the same way that surface animals seek out plants and algae. Although they're not reliant on the Sun for their food source, these creatures do need to take in oxygen, which comes from photosynthetic sources near the surface. Hence, only the anaerobic bacteria are truly independent from the Sun. To a hydrothermal Turner, the Earth's core would be God.

The discovery is remarkable enough in itself, but think of the implications. If life can thrive under such seemingly inhospitable conditions, then it only raises hopes of finding organisms elsewhere in the Solar System. Many scientists now think that life on Earth began at these hydrothermal vents. From the perspective of the chemosynthetic bacteria, it is us that live in the extreme environment, having evolved over billions of years to tolerate such nasties as sunlight and oxygen.

---

* FOOTNOTE This exceptional vehicle must rank alongside the Apollo 11 space capsule or the Hubble Space Telescope in its importance to exploration. As well as the discovery of hydrothermal vents, Alvin was also used to investigate the wreck of the RMS *Titanic*. It continues in service today, more than 50 years on from its first mission – though much upgraded and altered.

Fish were the first animals to leave the oceans

Ever seen a mudskipper? These peculiar creatures look like a cross between a frog and a fish, with protruding eyes and shiny skin. As the name implies, they can often be found skipping through the mud and sand of the seashore, using their front fins like legs. They are 100 per cent, *bona fide* fish, and yet they spend three-quarters of their lives out of water. They can even climb trees.

A creature similar to the mudskipper is thought to have been the ancestor of all four-legged vertebrates, including ourselves. At some point, a particularly plucky fish learnt how to venture onto land for short periods. Over time, its kind spent increasingly lengthy stays ashore. Gradually, gills gave way to lungs to produce the first amphibians, and thence reptiles, birds and mammals.

In 2004, the first well-preserved fossils of such a creature were unearthed in Canada. The so-called tiktaalik fish patrolled the world's oceans around 375 million years ago. It probably used its lobe-shaped fins to haul itself through shallow water and perhaps over land, before splashing back into the brine. Although a fish, it bears many of the hallmarks of later tetrapod (four-legged) species. Nobody knows if the tiktaalik is a direct ancestor of any species alive today – it is more likely to be an offshoot of the line that led to tetrapods, and it would therefore be misleading to call it a 'missing link' between sea and land. However, the fossils offer a revealing glimpse into the kind of creatures that might have bridged the gap between water-dwelling fish and land-dwelling animals.

Tiktaalik and its cousins were not, however, the first animals to emerge from the sea. The land was already teeming with life. The earliest hints – the supposed tracks of a scuttling arthropod species – date back to 500 million years ago. Much earlier than this would pre-date the formation of the ozone layer, exposing venturesome life forms to UV radiation. Our pioneering arthropod would have found a dusty, barren land. It may have encountered the odd patch of algae here and there, but land plants had yet to appear.

The oldest fossilized animal found so far is a millipede called *Pneumodesmus newmani*, which rippled over the dirt 428 million years ago. We know of this creature from a single fossil, found on the Scottish coast by bus driver Mike Newman in 2004. The artifact shows the presence of spiracles, the small openings used in respiration that only work out of water.

From such beginnings, myriad forms of life flourished on the land, including the earliest insects. Those crawling fish had plenty of company when they first flopped onto the land.

The meteor that killed the dinosaurs caused the biggest mass extinction of all time

If you ever gain access to a time machine, be careful not to set the dial to 66 million years BC. You really don't want to hang out there. A disaster of merciless magnitude befell the planet, one that killed off most of the large animals.

Imagine a lump of rock as wide as Paris, but without the pretty boulevards and pavement cafes. Now picture it ramming into the Earth at 108,000 km/h (67,000 mph). Had you been standing in the Gulf of Mexico all those millennia ago, this would have been the last thing you ever saw. The impact blasted a crater 177 km (110 miles) wide, and 19.3 km (12 miles) deep. Needless to say, this was a pretty bad day for anything on the same side of the planet.

Indeed, the whole planet would have experienced the shockwaves from such a strike, followed by further earthquakes and volcanic eruptions. This deadly medley would have kicked up choking clouds of dust that would have blocked sunlight, acidified the rain and impeded plant growth for years afterwards. Around three-quarters of plant and animal species were wiped out. The whole food chain was broken. Larger animals, including most of the dinosaurs, didn't stand a chance. The event was so significant that it marks the turning point of geological ages. The Mesozoic Era ended and the Cenozoic began; or, in more cuddly terms, the Age of Reptiles gave way to the Age of Mammals.

Scientists are now very confident that the cause of this mass extinction was an impact (probably an asteroid) in what is now the Yucatan Peninsula of Mexico. Sixty-six million years of erosion and infilling have largely deleted the crater, but its telltale curves are visible using certain geophysical techniques like gravity mapping. The dating of the impact coincides neatly with the last known dinosaur fossils. This timing is persuasive, although rival theories for the mass extinction still circulate, including the possibility of multiple impacts. Recent research suggests that the dinosaurs had been in decline for 50 million years before the knockout punch from the skies.

Whatever did for the dinosaurs and their friends, this extinction event is a mere fly swat compared with the so-called Permian-Triassic event of 252 million years ago. This was the bleakest time to be alive, especially for ocean dwellers. It's estimated that 96 per cent of all marine species were wiped out. On land, 70 per cent of vertebrates were smited. It is the single worst disaster to ever befall the Earth, yet we have no idea what caused it. The catastrophe was so long ago that the smoking gun has forever blown cold. Another asteroid impact is a possibility (other large craters are known), though it might equally have been caused by huge volcanic activity, a runaway greenhouse effect, a drop in oxygen levels or perhaps a combination of several environmental challenges.

The steamroller of extinction has trundled around the planet on several other occasions. An unknown catastrophe occurred around 450 million years ago that wiped out 70 per cent of all species. The dinosaurs themselves benefited from another event, a little over 200 million years ago, which killed off many of the competitor species and tipped the Triassic period into the Jurassic. There is no clear definition of what constitutes a 'mass extinction', but scientists tend to talk about a 'big five', and many minor events. The one that killed most of the dinosaurs is the most recent biggy, but by no means the biggest.

Some scientists believe we are now experiencing a sixth large event, often dubbed the Holocene extinction. By some measurements, species are currently dying off at a thousand times the rate estimated before 1900, mostly thanks to human activity. Including plants, corals and insects, one study put the number of extinctions at 140,000 species per year. Think about that. Almost 400 species disappear every single day. Can you even name that many organisms? This sudden rise in extinctions, coupled with pollution, habitat loss and other man-made interventions, has led many scientists and commentators to suggest that we have entered a new epoch, dubbed the Anthropocene.

Evolution is a slow process that takes many thousands of years

Our monkey-like ancestors took a very long time to evolve into modern humans. We only split off from chimps (or, more properly, a species that would later evolve into chimps as we know them) around 13 million years ago. Similarly, the farmyard cockerel is aeons removed from its feathered dinosaur forebear. When we think of evolution, we tend to picture the big stories and the big creatures. Most evolution is on a smaller scale, and is much more rapid.

Evolution by natural selection works by tiny changes in each new generation. Imagine you're a frog living in a lake that is gradually being choked by pollution. Most of your tadpoles will succumb to the pollution, but one or two will have a random distinction that will help them survive – maybe they carry a mutation in a protein that allows them to take in slightly more oxygen, or else they have more sensitive receptors for detecting pollution. These are the tiddlers that will survive into adulthood. Their own offspring may inherit the advantage and pass it on to their own offspring. With time, the population as a whole will share the advantage and become more tolerant to the pollution. That is natural selection.

Changes in appearance can take a very long time. Our example above would involve several generations of frog, and therefore several years – and this merely for a small tweak to a microscopic protein. To see a more radical transformation might require many thousands of generations.

All this means that evolution can seem to take forever when we look at larger creatures. But consider bacteria. They can multiply much faster than frogs. It takes as little as ten minutes for a bacterium to divide in two. Each of these can then divide again within another ten minutes. After an hour there would be 64. After a day, that one microbe could have become billions. This means we don't have to wait a long time for bacteria to mutate and change to suit a new environment.

Or a new drug. The rise of antibiotic-resistant drugs is one of the most serious challenges of our age. Frontline medicine is becoming increasingly ineffective, as bacteria learn how to shrug off our best antibiotics. If the decline continues, a simple cut to the finger may prove fatal. Surgery and childbirth will once again carry grave risk. The threat is upon us thanks to rapid evolution.

Antibiotics are potent drugs and will knock out almost all the bacteria they're designed to kill. Not all bacteria are created equal or identical, however. Each comes with random mutations in its DNA. These may have been introduced during replication, or by outside influences such as UV light. The vast, vast majority of these mutations will do nothing. There's a minute chance, however, that the right combination of mutations will occur in exactly the right bits of DNA to cause a shift in the bacteria's cell machinery, allowing it to block the drug. It is then a drug-resistant bacterium.

The chances of such a set of mutations in any given bacterium are vanishingly small. But we're dealing here with populations of billions and billions of bugs. The right combination is sure to arise sooner or later. If that blessed bacterium is then exposed to a run of antibiotics, it may well survive while its cousins die out. Within hours it may have divided billions of times. The host human won't be able to shake the infection. Without proper quarantine, the drug-resistant bacteria will spread to others, and cause an epidemic. It is a frightening and very real prospect. It is also an example of evolution happening at rapid speed*.

---

* FOOTNOTE This is a simplified picture to emphasize the role of evolution. Bacteria are also capable of swapping sections of DNA with their neighbours, including other bacterial species. Over time, some bacteria acquire the right DNA to resist several different antibiotics – like poker players holding a royal flush.

Nature never invented the wheel

The creatures of the Earth slither, scuttle, crawl and walk. They leap and prance; dart and dance. But they never roll. The wheel has been a transformative technology for humans. Without it, we could never shift anything much heavier than ourselves. Yet it's often observed that a wheel never evolved in nature. Nobody ever saw a gazelle on castors, nor a bison with an undercarriage. If millions of years of evolution can give us the brain and the eye, why not a chassis?

As it happens, wheel-like devices have been discovered in nature – it's just that they tend to exist in creatures we rarely examine close up. Indeed, one of the commonest methods of moving around on the planet uses a form of wheel. Many species of bacteria, perhaps half, sport a whip-like tail known as a flagellum. A special protein anchored in the cell membrane works like a speedboat's outboard motor, spinning the flagellum round and propelling the microbe forward. A similar system has evolved independently in the archaea – a branch of life similar to bacteria. A few other biological systems rotate, while not exactly resembling a wheel, including the enzyme ATP synthase and a rod known as a crystalline style, found in certain bivalves and gastropods.

And yet the question remains: how come we don't see wheels on larger animals? One obstacle may be evolution itself, which tends to work in small, incremental steps. We can readily imagine a pathway to the eyeball, for example. Some ancient creature developed a small number of cells sensitive to light, which gave it a slight advantage in avoiding approaching predators. Over millions of years, the number of light-sensing cells increased and diversified until eventually the creature had an organ akin to an eye. A wheel, by contrast, either works or it doesn't. It is difficult to imagine a first step to a wheel that might have any advantage over simpler ways of moving about*. Just how useful would a perfect set of wheels be, anyway? Wheels work best on flat, hard terrain, whereas much of the Earth's land surface is bumpy, and covered in dense vegetation or impassable sand dunes. Imagine climbing a tree or a cliff if your limbs ended in discs. Environments in which wheels might confer an advantage over legs are limited. Legs offer more versatility, one of the reasons that amputees opt for artificial limbs rather than unicycles.

A final difficulty is biological. Try to imagine a system of axles and wheels made out of flesh and bone – a kind of meaty wagon, if you will. It's hard to envisage such an anatomy that wouldn't be ground down by friction and infection. Even if we invoke a biological lubricant to grease things along, we have the problem of cabling. Such a dynamic piece of biology would need a sturdy blood supply. How do you attach the veins, arteries and nerves without risking a dangerous entanglement as the wheel turns? Of course, this too could just be a failure of imagination. The curious reader might pass an idle hour or two trying to find some workarounds, perhaps in insects.

Finally, certain types of animal behaviour mimic the wheel without actually using one. Picture, for example, a dung beetle riding along on top of its feculent prize. Hedgehogs, armadillos and pangolins all curl up into a ball when threatened, and may then roll a short distance to escape. We might not find a true wheel in larger animals, but the circle of life certainly throws up some analogues.

* FOOTNOTE We should, of course, be careful of such reasoning. Nature, in her infinite bounty, often throws up surprises that confound human imagination.

Humans are the pinnacle of evolution

Meet the ancestors.

Pictures like this, showing the rise of humans from monkey-like ancestors to spear-carrying hunters, are common currency, to the point of parody. We've all seen the alternative versions, where the monkey–ape–caveman sequence terminates with a figure hunched over a keyboard or smartphone, as though to say 'so this is the crowning achievement of evolution'.

That notion that humans represent the pinnacle of evolution is fallacious, and it is often debunked in popular science books. We are, by many measures, an impressive species. You don't see whelks assembling a moon rocket, nor badgers sending text messages. But think about it for a second: we are only a pinnacle under our own terms. Those most recently maligned whelks might not have a space program, but they are much, much better than humans at clinging on to rocks and breathing under water. In their own sub-aquatic environment, they are closer to a pinnacle than our species. Stick a human and an anteater on a termite mound. One will thrive, the other will scream.

Supremacy also depends on your definition of success. Population size might be one measure. After all, the primary goal of all life forms is to reproduce and pass on DNA to the next generation. By this yardstick, humans are not even pitiful. Our species numbers some seven billion individuals. Nobody has ever counted all the ants on Earth, but estimates stray into the trillions. Argentine ants alone exist in super-colonies of billions of individuals.

This is just to ease you in gently. If we move on to bacteria, the numbers are extraordinary. Our best guess puts their tally in the region of 5,000,000,000,000,000,000,000,000,000,000. Name any particular species of bacteria and it would outnumber humans many millions of times over. Your left nostril probably contains more of these life forms than the sum total of humans who have ever lived. In comparison, we are so woefully non-fecund that our existence is negligible. Pinnacle, schminnacle.

Another measure of success might be longevity. Our planet, as much as we love it, is full of hazards. Any species that can survive unchanged through the millennia without getting wiped out must be doing something right. Here the jury is still out on *Homo sapiens*. Modern humans have walked the Earth for a mere 200,000 years or so. Crocodiles, by contrast, have been snapping away for 50 million years. Stromatolites – a type of bacteria that forms into rock-like colonies – have remained physically unchanged for a billion years. We are mere infants.

Even individual organisms may be older than our entire species. In 2009, microbiologist Raul Cano made headlines after brewing ale with yeast that was 25–45 million years old. The microorganisms were extracted from an ancient block of amber – much as in *Jurassic Park*, but with fewer claws and more hangovers. The yeast had lain dormant through aeons, surviving ice ages, the motions of tectonic plates and the rise of human civilization. Now, revived in Cano's brewing vats, the yeast is turning sugar into alcohol so we can all get drunk. The beer, since you ask, is said to taste spicy, with a hint of cloves.

The notion that humans represent an evolutionary peak is a subjective one. We might have the most sophisticated brains and an unrivalled control of our surroundings but, reframed, we're lesser creatures than whelks, germs and brewer's yeast.

The Latin name for humans is Homo sapiens

Any keen gardener will know that his or her plants have fancy Latin names. Heather, for example, is more soberly known as *Calluna vulgaris*, while common holly is *Ilex aquifolium*. The system is called binomial nomenclature. It bequeaths a double name on every known plant, and these tell you both the plant's genus and its species. *Quercus robur*, for example, is just one of many species of oak tree that sits within the genus *Quercus*. You might also be aware of the Turkey oak (*Quercus cerris*), the holly oak (*Quercus ilex*), or any of their 600 other acorn-bearing cousins.

Such binomial nomenclature was first advanced by Carl Linnaeus in 1735. He applied the system* not just to plants, but also to animals and minerals (the parlour game '*Animal, Vegetable, Mineral*' was a whimsical spin-off from his insight). Linnaeus gave the name *Homo sapiens* to our species, and we still use this today. It translates, rather flatteringly, as 'wise man' or 'sapient man'.

---

* FOOTNOTE Linnaeus didn't stop with species and genus in his system of classification – he nested these names under still higher categories of life. In his scheme, every genus was part of a family, every family was a member of an order, every order belonged to a class, and classes ultimately group into one of those three kingdoms from our parlour game. The system has been much modified over the centuries, and now includes many further groupings. Even the all-powerful kingdoms (animal, vegetable or mineral) have been redefined, added to, and slotted under even higher taxonomic categories known as domains. Minerals, not being a form of life, are no longer part of the system.

But here's the error, often made: *Homo sapiens* is not the Latin name for humans. It is the Latin name for just one species of human – us. The two terms are often used synonymously, because we never encounter other species that call themselves human. However, humans were happily going about their business millennia before *Homo sapiens* appeared.

Confusing? Let's back up a bit. We've all encountered illustrations of our distant 'apeman' ancestors. They look a lot like us. More hair, perhaps, with a stooped posture, simian faces and spears instead of smartphones – but very much human in other respects. Indeed, many of them were human. The term is used not just for our own kind, but also for the many earlier bipeds from the genus *Homo*. Now-extinct forebears, such as *Homo habilis, Homo neanderthalensis* and *Homo erectus,* are all considered to be species of human, in exactly the same way that *Quercus robur, Quercus cerris* and *Quercus ilex* are all species of oak tree.

Homo sapiens that look just like you or me have been around for something like 200,000 years. The earliest signs of a culture, including artistic and religious relics, date from around 50,000 years ago. Our species from this point on is often termed 'Modern man' or *Homo sapiens sapiens*. Humans, on the other hand, stretch back much further. *Homo habilis,* the earliest known member of the genus *Homo,* debuted almost three million years ago. The *habilis* species lasted well over a million years – much longer than we have so far managed. In terms of biological longevity, modern humans are a minor player in the pageant of humanity.

Bigfoot nonsense aside, only one species of human exists today, and that's us. Imagine a different world, though, in which we share our land with stocky Neanderthals and metre-high dwarf humans. One would only have to travel back a few thousand years to encounter other flavours of human. Neanderthals are thought to have died out around 40,000 years ago. They argued, fought and slept with our *Homo sapiens* ancestors (see next section). The Hobbit-like *Homo floresiensis* of Indonesia, meanwhile, may have survived until as recently as 12,000 years ago – although new evidence suggests a more likely date of 50,000 years. To equate the term 'human' solely with *Homo sapiens* is tantamount to racism, although nobody's around to take offence.

So now we're clear what *Homo sapiens* is and what it isn't. But be sure to get the spelling right. It's tempting to think 'sapiens' is a plural, which would mean an individual of the species would be a *Homo sapien*. That would be wrong. Sapiens, as we've already seen, means 'wise'. It is an adjective. The 's' needs to be on the end, and there is no such word as *sapien*. So if you're talking about a lone representative of our species you should still say *Homo sapiens*. For example: 'Lassie is a *Canis familiaris*, while her owner is a *Homo sapiens*.' Also note that genus and species names are, by convention, always italicized. Genus names get a capital letter (*Homo, Quercus*), but species names do not (*sapiens, robur*). And you can abbreviate the genus, so long as you add a dot: (*H. sapiens, Q. Robur*). These rules are flouted as often as they're obeyed.

And finally, back to Linnaeus. The Swedish botanist not only named our species, he also became the quintessential modern human. Scientists like to designate a 'type specimen' of every species – that is, a specimen that serves as the gold standard and official embodiment of that species. Guess what? Linnaeus is the type specimen of modern humans. Unlike many types, however, Linnaeus's remains do not reside in a museum or archive, but in a tomb at Uppsala Cathedral.

Modern humans are descended from Neanderthals

Of all the human species, Neanderthals hold the strongest appeal in the imagination. It may be their distinctive appearance – hairy, with heavy brows and big noses – and the persistent use of 'Neanderthal' to describe someone who is slow or gruntish. As we saw in the previous section, however, Neanderthals are just one of many human species that once walked the planet.

A common misconception – reinforced by those clichéd drawings of knuckle-dragging apes sequentially evolving into upright hunters, then office workers – is to assume that we modern humans are descended directly from the Neanderthals. It's a bit like saying that modern elephants are descended from the woolly mammoth. Neither idea is true. Neanderthals lived alongside modern humans and both descended from a common ancestor, somewhere between 400,000 and 700,000 years ago, and probably in Africa.

But that's not the whole story. Recently, scientists have discovered that our modern human genome is not entirely pure. Some 2–4 per cent of non-African DNA (it varies from person to person) matches that of Neanderthals. There's only one conclusion: modern humans and Neanderthals had plenty of sex with each other. The last of their kind died out 40,000 years ago, but a tincture of Neanderthal lives on within our cells.

Since that discovery, scientists have identified many stretches of DNA that can be traced back to our cousins. Genes contributing to paler skin, freckling, depression and a greater risk of Type 2 diabetes, for example, have all been linked to a Neanderthal past, although the research is still at an early stage. One estimate suggests that 40 per cent of the Neanderthal genome is still 'out there', chopped into pieces and circulating piecemeal among modern humans. In one sense, then, the heading of this section is true – we have inherited much from the Neanderthals, though can claim no direct descent.

By the way, the cliché of the dim-witted Neanderthal is also of shaky accuracy. Our cousins had a basic knowledge of cookery and medicine. They even built sculptural rings of stalagmites, suggesting they were capable of symbolic thought, and possibly art.

Our promiscuous ancestors didn't restrict their exotic couplings to Neanderthals. Another branch of humanity known, less catchily, as the Denisovans also interbred with our ancestors. They too have left their mark on our genomes, particularly among the people of south-east Asia. In case you're wondering, Denisovans and Neanderthals also bunked together, raising the prospect of the ultimate love triangle for any scriptwriter working on a Paleolithic soap opera. Actually, wouldn't that be a good idea?

Scientific misnomers and misquotes

Arguments over nomenclature and definition are a common occurrence in science. Here are a few of the less technical examples.

Aluminium versus aluminum The rule seems clear-cut. All the world calls this metallic element aluminium, except for those in North America, who insist on dropping the second 'i' to forge aluminum. But not so fast, for the American spelling was first coined in Britain.

Chemist Humphry Davy never managed to isolate aluminium, but he did bestow upon it a name. Two names, actually. He first dubbed the substance alumina and later settled on aluminum. Others objected, favouring aluminium in sympathy with potassium, calcium and magnesium. Still, Davy's alternative ending took root in the USA, finding its way into *Webster's Dictionary* in 1828. The spelling was bolstered later that century by baby-faced aluminium magnate Charles Martin Hall. The American inventor and entrepreneur was among the first to open large-scale aluminum production plants. Despite plumping for aluminium in his patents, Hall later used aluminum for all his products, ensuring the name variant became the standard in his territory.

Today, both terms are commonly used, in both general speech and scientific communication. The International Union of Pure and Applied Chemistry (IUPAC), the body responsible for scientific nomenclature, lists aluminium as the international spelling, but allows aluminum as a common alternative. It's certainly a popular 'alternative'. Google Scholar, a search engine for scientific literature, lists almost twice as many citations for aluminum as aluminium. Maybe we should call it a draw and all switch to Davy's first suggestion of alumina.

Asteroid The name, coined by Uranus-discoverer William Herschel, is of Greek derivation and means 'star-like'. Asteroids could hardly be less star-like, however. They are cold, irregularly shaped lumps of rock that orbit in their thousands around the Sun (with a particular concentration between Mars and Jupiter). Through his primitive telescope, the objects were too small for Herschel to descry any detail. He saw only pinpricks of light, and so compared them to the distant stars.

Dark side of the Moon It might not look like it when seen from Earth, but the Moon is spinning. It rotates at precisely the right speed to keep the same face towards Earth. If this seems baffling, try an experiment with your fingers. Point inwards with your left index finger. Now, slowly rotate your right index finger around your left. You'll notice that your right finger shows an ever-changing face to the left. If you wanted to keep your nail forever pointing at your left finger, you'd have to rotate your right hand (which, due to the limitations of human anatomy, is not a possibility).

Because we always see the same face of the Moon, it follows that there is another side that we never see, at least from Earth. This is traditionally called the 'dark side', as popularized in the famous album by Pink Floyd. It's something of a misnomer as the far side of the Moon gets as much sunlight as the side that we see, unless 'dark' is taken to mean 'mysterious' or 'occluded'. No one knew what the 'dark side' looked like until a space probe passed by in 1959.

Dinosaur The name means 'terrible lizard'. The trouble is that dinosaurs were not lizards but an entirely separate group of reptiles. It is thought that dinosaurs had greater internal control over their body temperatures than lizards, though they were neither 'warm-blooded' nor 'cold-blooded' (themselves misleading terms), but somewhere in between. In addition, many creatures we think of as dinosaurs – pterosaurs, pleisiosaurs and dimetrodon among them – are nothing of the kind. Meanwhile, birds are today considered to be a specialized subgroup of dinosaurs, which means the dinosaurs never did die out completely.

Guinea pig Not a pig; and not from Guinea.

Halley's Comet The Solar System's most famous comet is named after Edmond Halley, the Astronomer Royal, so you'd be forgiven for thinking he discovered it.

Think again. The comet is easily visible with the naked eye and has been noted since ancient times (most iconically on the Bayeux Tapestry, recording the Norman Conquest of England in 1066, which followed the comet's appearance). Halley, working in 1705, was the first to spot its periodicity of 75–76 years using accounts from earlier documents.

Meteor/meteorite/meteoroid Effectively, these are three distinct stages of a small rock or natural metallic object as it moves toward Earth. When travelling through space, the object is known as a meteoroid. When the meteor enters the Earth's atmosphere it is termed a meteor. It is the ephemeral streak of a meteor that we call a shooting star. Most of the object is obliterated by its passage through the atmosphere. Anything that reaches the ground is called a meteorite.

Royal Society The British have a singularly creative history of inventing organizations with unhelpful names – especially so in the sciences. Take the Royal Society. Founded by the likes of Christopher Wren and Isaac Newton in the middle of the 17th century, it is one of the world's most venerated and venerable learned societies, with a magnificent HQ overlooking The Mall in London. But you can't help thinking: Royal Society of *what*?

The name would be only mildly confusing were it not for the existence of a second estimable scientific organization just half a mile away in Mayfair. The Royal Institution (often mistakenly called the Royal Institute) is a relative youngster, founded around the turn of the 19th century. Again, the name raises the question: Royal Institution of *what*?*

Both have an excellent track record of putting on public lectures and demonstrations; both probably have an excellent track record of receiving each other's post and each other's visitors. A third body, the British Association for the Advancement of Science, once revelled in similar acts of obfuscation. It was

---

* FOOTNOTE Technically, it's the Royal Institution of Great Britain, but that says nothing about what the organization actually does, and nobody calls it by this full name.

known by almost everyone as the British Association, as though shy of its role in advancing science. A wise rebrand in 2009 changed all that, with the creation of the British Science Association.

Stone Age A non-scientific term, often used to denote the period before humans had learned to craft metals. It should really be called the Wood Age. Trees would have yielded far more tools than rocks to those ancient people. We have far more examples of their rock technology though, simply because wood tends to perish while flint axe-heads last forever.

Sulfur/Sulphur Debate can get rather heated over this one, which is appropriate for an element associated with fire and brimstone. It's often assumed that element number 16 is spelled with an 'f' in American English and a 'ph' in British English, and this was indeed true for much of the 20th century. Since 1990, however, the IUPAC has advocated 'sulfur' as the international standard. The Royal Society of Chemistry followed suit in 1992, giving a British nod to the simpler form. Of course, there was an outcry. It continues to this day whenever a British publication dares to write 'sulfur'. Yet there's little other than tradition to back up such obstinance. The 'ph' spelling is common in words loaned from Greek, as a way to replace the letter 'phi'. Unfortunately, sulfur is ultimately derived from Latin, not Greek. Sulphur is trumped by sulfur on arguments of both etymology and authority, yet many refuse to let it go.

Theory In everyday language, a theory is little more than a hunch. 'I have a theory,' you might say, 'that Adam Sandler is the greatest actor of this or any other age'. Or if someone asks whether you're free for lunch next Tuesday, you might say 'Yes, in theory'. It is a lightweight word, used with flippancy. To a scientist, on the other hand, a theory is a serious thing.

Compare your theory about Adam Sandler with Einstein's Theory of General Relativity. The first is your opinion. It is subjective and unverifiable – unless you invoke a measure such as Oscar nominations, which in this case would be a fool's gambit. The latter is built on centuries of previous work by dozens of scientists. General relativity has remained the consensus view of gravitation for over a century. Countless experiments have confirmed its predictions.

And yet we still call it a theory and not a fact. This point is at the very heart of science. No matter how established an idea gets, it is always open to question. Scientists will merrily chip away at the corners of any big theory, in an effort to find exceptions or inconsistencies. That's where progress lies. Newton's theory of gravitation held for hundreds of years. It's still a superb tool for everyday Earth-bound stuff, like calculating the flight of a tennis ball. But we now know, thanks to Einstein, that there's more to gravity than the three laws of motion.

Vitamin D Strictly speaking, a vitamin is an organic compound necessary for health that is not synthesized in the body and must be gained through diet. Vitamin D is an oddity in this respect. It's readily made in the body so long as that body is getting plenty of sunlight. However, most people get more than enough from foods, such as eggs, mushrooms and fish.

Planet Earth

To quote Carl Sagan,
'On it everyone you love, everyone you know, everyone you ever heard of,
every human being who ever was, lived out their lives'.
But how well do we know planet Earth?

The last ice age ended thousands of years ago

Look out of the window. Do you see any ice sheets or towering glaciers? I'm guessing the answer is 'No', unless my publisher has been exploring some highly exotic new markets. It might have been a different matter 20,000 years ago. The so-called Last Glacial Maximum saw all of northern Europe covered in ice. In North America, the glaciers ground down as far as Manhattan. The Alps, Andes and Himalayas were largely hidden beneath the white stuff, while the rest of the planet was cooler and drier than now. With so much water locked up in polar ice, global sea levels were considerably lower, creating land bridges between areas that before and after would be separated by seas. It was a very different world.

And then the ice began to recede. Sea levels rose once more, and the climate warmed. Approximately 12,000 years ago, the polar ice caps had retreated to a point that would look familiar today. The long ice age, which had lasted around 100,000 years, was finally over.

Or was it? It all depends on your definition of the term 'ice age'. Most scientists would tell you that we're still in the midst of one.

Although today's climate is warmer than the frigid conditions we associate with woolly mammoths and hunters in bear skins, we still have significant ice cover around the north and south poles. This is unusual. Over the long lifetime of our planet, the normal state of affairs is for an ice-free world. It just so happens that the northern- and southern-most latitudes have been frozen for the recorded history of our species, so we naturally think that ice caps are normal. They're not. Just four other icy periods have been identified in the entire 4.5-billion-year history of the Earth. These are the ice ages.

The march of the glaciers that lasted until 12,000 years ago is – almost literally – the tip of the iceberg. The ice has been growing and retreating regularly over the past 2.5 million years. We're currently in an 'interglacial', a brief warm period within an ongoing ice age. While we still have icecaps at the poles, we are technically still in that ice age.

Nobody knows how long the ice sheets will yo-yo before disappearing completely and ending the current cycle. It could be a long time yet. The most impressive example, known as the Huronian glaciation, spanned 300 million years – in other words, a cold spell that lasted 1,500 times longer than the existence of our species.

It would be tempting to think that man-made global warming might tip us out of the cycle and prevent a return of the glaciers for a long time; maybe, maybe not. Recent research suggests that the next glaciation might have been delayed by up to 50,000 years, thanks to the amount of heat-trapping CO_2 we've released into the atmosphere. On the other hand, further melting of the ice caps might play havoc with ocean currents such as the Gulf Stream, potentially tipping us back toward a frozen north.

Earthquakes are measured on the Richter Scale

We all know the Richter Scale. The system for categorizing earthquakes is so famous, it's even namechecked in songs by the Beastie Boys, AC/DC and Frank Zappa. Strange to report, then, that the Richter Scale is rarely, if ever, used by scientists.

The Richter Scale was devised in the 1930s by seismologists Charles Richter and Beno Gutenberg (so one could argue it should be the Gutenberg–Richter scale, particularly as Gutenberg was Richter's mentor). The duo had a specific purpose for their scale. They wanted to compare and categorize the energy released by earthquakes in just one area of California using one type of seismograph. Their system worked well under those conditions, but was unsuited to other locations. It was also inaccurate at larger magnitudes.

A new system, applicable anywhere in the world, was developed in the 1970s. The moment magnitude scale is now the standard used by all seismologists. When you hear a newsreader talking about an earthquake measuring eight on the Richter Scale, he or she is almost certainly mistaken. That said, the scales are not wildly different from one another. A magnitude 8 quake is bad news whichever system is used.

Two other misconceptions about the Richter Scale persist. It does not measure the amount of damage caused at the surface, but the amount of energy released. Later interpretations have added descriptions such as 'moderate to severe damage' and 'some objects may fall off shelves'. These make the scale more tangible, but can be misleading. A magnitude 6 earthquake, for example, might be more destructive than a magnitude 7, if it occurs much closer to the surface. Second, it's important to remember that both the Richter and the moment magnitude scales are logarithmic. An increase of magnitude 1 indicates a ten-fold increase in the seismograph reading (it's measured by the amplitude of the seismic waves). This means, for example, that a magnitude 8 quake has ten times the shaking amplitude of a magnitude 7 quake.

Water drains anti-clockwise in the Northern Hemisphere and clockwise in the Southern

Pull the plug in your sink and, after a few seconds, you'll notice that the water curls around the hole before disappearing forever. An old yarn would have it the water always spins clockwise if you're somewhere south, like Australia, and anti-clockwise for those in more northern latitudes.

The notion has a strong ring of truth about it. To see why, grab yourself an apple* and stick paper clips into the skin near the two 'poles'. Fold out the ends of the clips so they look like little flags. Now, slowly rotate the apple, left to right to mimic the Earth's spin. You'll see the northern flag is turning in an anti-clockwise direction when viewed from the apple's equator, while its southern counterpart runs clockwise. Scale things up to the size of a planet and, so the theory goes, you'll see this mapped onto draining water, which will curl in different directions in opposite hemispheres.

---

* FOOTNOTE This would work with oranges, too, but don't blame me if the juice squirts in your eye.

The theory is sound, but not on the scale of the kitchen sink. The direction the water spins is much more heavily affected by the shape of the basin, existing movement in the water and any bumps, imperfections or encrusted toothpaste than it is by the spin of the Earth. You would need ideal conditions, with a very large, perfectly even bowl, to observe the difference.

The story is based on real science called the Coriolis effect. The spinning of the planet might not affect your bathtub, but it does play games with the oceans and winds. Remember that a position on the equator moves faster around the Earth's axis than a position anywhere else on the planet, just as children on the outside of a roundabout spin faster than the smug child sat in the centre (in other words, they cover more distance than the central child over any given time interval). By the same token, the atmosphere at the equator has more momentum from the rotation of the Earth than it would in other locations. This leads to differences in pressure and the rise of tropical storms. When viewed by satellite imagery, these spin anti-clockwise in the northern hemisphere and clockwise in the southern hemisphere. This is the so-called Coriolis effect, and it also drives ocean currents.

If you don't believe me about the draining sink, there's an obvious home experiment waiting to be done. Give your kitchen sink a good clean, fill it with water, then pull the plug. Which way does the water spin as it drains? Repeat this several times, and with different sinks. You'll probably get mixed results, whichever hemisphere you're in. At the very least, you'll have a nice, shiny basin.

Everest is the world's tallest mountain

The one thing every schoolchild learns about geography is that Mount Everest is the world's tallest mountain. As facts go, it's as unassailable as the mountain would itself be to the typical schoolchild. Yet, like so many other items in this book, it depends how you measure and define.

Everest's height is usually given as 8,848 m (29,029 ft). That measurement is the height of its summit above sea level. But what if a mountain carries on beneath sea level? That's the case for Mauna Kea, a volcano that crowns the main island of Hawaii. Take a look on Google Earth and it's clear that much of this volcano is beneath the waves. Measured from the seabed to its summit, Mauna Kea would top 10,200 m (33,465 ft). Were the oceans ever to recede, then this giant would, by all measures, be the tallest peak on Earth.

Other mountains can also stake a claim. If we were to measure the distance from the centre of the Earth to the tip of the summit, then Everest would be supplanted by the Ecuadorian volcano of Chimborazo. It stands 'just' 6,263 m (20,548 ft) above sea level but, because the planet bulges close to the equator, its peak is further from the centre of the Earth than Everest's by some 2,000 m (6,560 ft). The Peruvian mountain of Huascarán would also top Everest for the same reason.

A final candidate might be the Alaskan peak of Denali, formerly known as Mount McKinley. It rises 6,190 m (20,308 ft) above sea level, seemingly trivial compared with Everest. However, Denali's base-to-peak height is actually greater than Everest, as the latter mountain rests on top of the Tibetan Plateau, which gives its base a hefty leg-up. In conclusion, then, Everest is only the tallest mountain by the somewhat arbitrary standard of height above sea level.

The more intrepid breed of mountaineer could find still loftier peaks on other worlds. Everest is a mere foothill compared with Olympus Mons on Mars, once thought to be the tallest mountain in the solar system at 21.9 km (14 miles) high, and at least three other Martian mounds climb higher than any Earthbound prominence. The record for the tallest rise is now thought to be on the asteroid Vesta. Here, the central peak of an impact crater known as Rheasilvia climbs to 22 km, marginally besting Olympus Mons. Various other bodies have mountains bigger than Everest and Mauna Kea – from Io to Mimas to Venus, we find ascents to beat those on Earth.

While we're on a geographical mythbusting trip, it might reasonably be claimed that the Sahara is not the biggest desert on Earth. The most common definition of a desert is a barren area of land with little rain or plant life. We automatically think of sandy dunes and blistering heat, but it needn't be so. The vast expanse of Antarctica also fits the bill, and the continent is usually considered a desert, half as big again as the Sahara.

Rainbows contain seven colours

We all have our little rhymes for remembering the order of colours in a rainbow. British children might learn that Richard of York Gave Battle in Vain (red, orange, yellow, green, blue, indigo, violet). Or perhaps we recall that mysterious individual by the name of Roy G. Biv – I always imagined him to be some kind of severe American colonel. Such mnemonics are useful as far as they go. They help us recall the basic spectrum, which runs from red through to violet. But it is, of course, a simplistic picture of the rainbow, which offers a blurred continuum of hues rather than discrete bands. The traditional seven colours are somewhat arbitrary. We might as readily identify eight colours, or 32, or 12.

Isaac Newton was one of the first to experiment with refraction – the bending of light that, in nature, leads to rainbows. He developed prisms that could split white light into its component parts. The ensuing spectrum might be considered an artificial rainbow, though sharper and more saturated. Newton believed he could split light into five colours in this way – red, yellow, green, blue and violet. He later added orange and indigo to give the seven 'canonical' rainbow colours we still learn at school. This was not some whimsical decision. Newton sought harmony with other systems, such as the number of notes in a musical scale, the number of days in the week and the number of 'classical planets', visible since antiquity to the naked eye (Sun, Moon, Mercury, Venus, Mars, Jupiter, Saturn).

Scientists reckon that the human eye can perceive around 100 distinct colours in a spectrum. This is under ideal conditions, when the white light comes from a single point source, and is split cleanly into different wavelengths by a prism. A rainbow is a different beast. The light source here is the Sun, which is effectively a disc of light rather than a point. And then the scattering agent is not some finely honed prism, but millions of drops of water, which we call rain. Sunlight entering those drops bounces around (internal reflection) or is scattered at certain angles (refraction). A single raindrop will bat just a very narrow range of wavelengths – effectively one colour – to your eye, reinforced by its near neighbours. Drops higher or lower will be at the right angle to send you other colours. In this way, we perceive a series of hues in an arc across the sky – a rainbow.

We all see a slightly different rainbow, too. Even if you and I stood right next to one other, you would see light that was chopped up by different raindrops to the ones that spat their light at me. How many colours we see is really down to perception and the sensitivity of our eyes. We will all receive millions of different wavelengths from the scattered and reflected light, but our eyes and brains will simplify the picture to a smear of colour.

There are other 'colours' hidden in the rainbow, too, that no natural human eye will ever observe. We can imagine that the light reaching our eyes travels along as a wave. Light with the longest waves, we perceive as red, while the shorter wavelengths are the domain of indigo and violet. Light can have still longer or shorter wavelengths, but our eyes are not kitted out to see it. With the right detectors, we might see infrared bands above the rainbow, and ultraviolet arcs beneath.

And then there's pink. Have you ever seen it in a rainbow or a spectrum? You will look in vain, for there is no single wavelength corresponding to this colour. Some people argue that magenta (to give it the proper name) should not be considered a colour at all. Humans can perceive the colour because of the way our eyes work. Magenta is conjured up by the brain to interpret a situation where the eye receives equal amounts of red and blue light; that is, the wavelengths at opposite ends of the spectrum. Our eyes contain three types of light receptors called cones: one type specializes in red light, one in green and one in blue. So if we take in light with red and blue wavelengths, but nothing from the green part of the spectrum, then the eye sums the inputs from both the red and blue cones. The result is magenta and you see it thanks to a trick of the brain.

So the next time a shopkeeper tells you that she has that dress in every colour of the rainbow, ask her if she also does magenta.

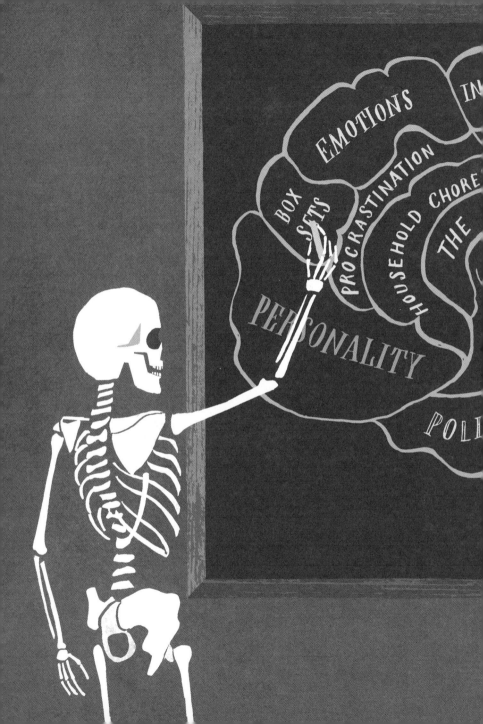

Body matters

Think you know yourself? Much of your body is not your own,
and the bits that are don't always function as you'd expect.

You are a human being

I'm making an assumption here, dear reader, but you probably think of yourself as a human being. In one sense, and frankly, the most important sense, you are. But appearances can be deceiving. About half the cells in your body are not human.

The rogue agents are bacteria, fungi and archaea. Your body is teeming with these tiny intruders. Anywhere between 500 and 1,000 species have made a home within your folds, ducts, flaps and chambers, and they're each present in their billions. It doesn't matter how strict your personal hygiene. The proportion is now thought to be roughly 50:50 human cells to microbes, though it may swing either way depending on your individual make-up, and whether you've recently visited the toilet.

Known as the human microbiota*, this collection of hitchhikers is mostly harmless and, in some cases, beneficial. Several species in the gut, for example, aid digestion, while others produce useful molecules for which our own bodies lack the blueprint. Such is their contribution that some have dubbed the microbiota a 'forgotten organ'. Even so, much has still to be learnt about the roles individual species might play.

The number of microorganisms in your body has itself become the subject of a popular misconception. It is frequently written that bacterial cells in the human body outnumber our own cells 10:1. This startling conclusion was recently debunked with more careful measurement and extrapolation. Scientists now

---

* FOOTNOTE Formerly known as microflora, which is a misnomer since 'flora' should strictly only refer to plant species.

believe the ratio is more like 1.3:1 in a typical person, though with some room for variation. You are half human, at least. When looked at another way, our own cells are much larger than those of microorganisms, so even with a 10:1 superabundance of alien cells, most of your body weight (excluding water) comes from your own stuff.

Yet another foreign contingent lurks within. As a placental mammal (again, I'm presuming), you spent your first nine-or-so months exchanging fluids with your mother. Something tangible often remains of that bond, long after birth. Many mothers retain cells from their babies in a process known as microchimerism. (The phenomenon gets its name from the Greek legend of the chimera, a strikingly awkward beast that was part lion, part goat and part snake.)

The child's cells don't just hang about inside the mother. They find their ways to different corners of her body, where they quickly adapt to match the local tissue type. For example, cells might reach the heart, latch on and turn into cardiac tissue. Somehow, this transformation helps the intruder cells evade the mother's immune system. The masquerading cells can remain here for several decades, functioning and dividing alongside the mother's native cells, yet genetically distinct.

If the mother then goes on to have further children, or even late-stage miscarriages, she's likely to take in further foetal cells. Poor Queen Anne of Great Britain (1655–1714) gave birth 17 times. None of the children survived for very long, yet her frequent gestations could have given Anne a cell population from as many as 18 different individuals. Talk about royal inbreeding.

It gets weirder. Less frequently, cells from the mother pass into the foetus. This being the case, the baby might also take in cells from its mother's previous pregnancies. You might carry fragments of an elder brother or sister. A child who died in infancy might live in remnant, its cells hidden among the tissues of its mother and siblings. As the cells can persist for decades, they might even be passed on to a further generation. Cells from your grandmother might loiter in your abdomen; a tincture of your uncle may sequester in your spleen.

Research into microchimerism is still at an early stage. It is not known how these cell swaps might affect the recipient, positively or otherwise. Some scientists think that the foetal cells – which have the ability to transform into any cell type – could play a part in repairing tissues in the mother's body. Others have suggested that the cells act in the interests of the foetus, delaying the onset of further pregnancies. Still others propose a link to autoimmune disease. Conversely, it's possible the cells have no effect whatsoever. In the cliché of science writing, 'more research is needed'.

Your body, then, is neither wholly human nor exclusively the preserve of one human. When Prince Hamlet mused 'What a piece of work is man!', a scientifically clued-up Rosencrantz might have replied: 'Why sire, a corporeal collection of hundreds of species, most of which are invisible and not human, with additional contributions from your sisters, brothers, father, mother, aunts and uncles; this family jumble is, in your case, my Prince, richly ironic'.

Hair and nails continue to grow after death

The notion that hair and nails continue to grow beyond death is a common one, perpetuated in Gothic horror stories, a thousand B-movies and, most recently, *The Walking Dead*. Is it true? Rigorous clinical trials would have a hard time winning ethical approval, and no one has done a controlled study, but nobody needs to. The idea of any part of the body continuing to grow after the heart has stopped is nonsensical.

For any tissue to grow, its cells must divide and multiply. This needs energy in the form of glucose, with a side serving of oxygen to drive the process. If your heart isn't beating, then your blood isn't circulating; if your blood isn't circulating, then no fresh oxygen is reaching your extremities. Without this oxygen, cells can't divide; your hair and nails cannot grow.

The misconception has a simple explanation. Not long after death, the body starts to lose moisture. The skin dries out; it shrivels. Hair follicles protrude more noticeably against this withered backdrop. The effect is most marked in stubble. As the skin recedes, the clipped hairs stand out more, giving the appearance of growth after death.

The myth might also be connected to rates of decomposition. Both hair and nails are formed from keratin, a tough fibrous material that can take centuries to break down. Exhumed corpses often retain traces of hair and nail after other tissues have rotted away. They haven't grown, they just haven't perished.

Artistic, creative people use the right side of their brains, while more analytical types use the left brain

Give a small child an abacus and she will behave in one of two ways. She might try to count the beads and perhaps make simple calculations; alternatively she might rattle it around, admire the colours and smash it over the head of a nearby playmate. 'Don't worry, she's just exercising her right brain,' her embarrassed parent might explain.

It's a common belief that our personality can be split between the two sides of the brain, with arty skills in the right hemisphere and more analytical skills on the left. If you look at a rock and imagine carving out the Venus de Milo then you're doing some serious right-brain thinking. If the same block of marble has you contemplating the age of the Earth, the sedimentary layering of rock and the high temperatures and pressures needed for metamorphosis into marble, then you're probably a left-brain thinker.

It's nothing like as simple as that. The human brain is too complex, too interlinked to be so crudely generalized. The idea that van Gogh made more use of his right brain, while Newton had an overdeveloped left brain, is pure bunkum. Such broad statements are a staple of pop-psychology magazines, but have no grounding in reality.

It's true that certain mental tasks are performed predominantly in one brain area. For example, most aspects of speech are controlled from the left hemisphere, while the right side handles much of facial recognition. But your personality type cannot be pegged to a particular hemisphere. Imaging experiments have shown that individuals do not typically favour one side of the brain over the other. Rather, the two hemispheres work together on most tasks.

This makes sense when you think about it. Imagine again that you're carving the Venus de Milo. Your brain needs to exercise precise control over your arm and hand muscles. You must calculate the best angles and most appropriate force with which to strike the marble. You need to use your imagination to picture the finished figure, and translate this into a spatial reality. You probably also have a motivation for making the sculpture in the first place, and a sense of the Venus's importance in the history of art. You weighed up the cost of your training and materials, and crafted the perfect 140-character tweet to show off your creation. Your sculpture is not simply an act of creativity; it demands many skills from its maker, some of which are sparked from the left brain, some from the right, but resulting most from an interplay of both.

The brain is made up entirely of neurons

The body contains dozens of cell types. Some are well known, like the red blood cells that carry oxygen round the body, or the skin cells, which are shed to become household dust. The brain is also made up of cells. One well-known cell type, called the neuron, is responsible for passing electrical signals around the brain and throughout the nervous system. But this is only part of the story.

Glial cells get much less attention, but probably outnumber neurons in the brain. These cells do not carry nerve impulses but play a number of support roles. Some provide the neurons with nutrients and oxygen; others serve as cleaners, removing toxins and dead cells. Still others fulfil a structural role, holding the neurons in a particular orientation. Others wrap around the neuron to form an insulating sheath. No neuroscientist has yet convinced an intern to do an accurate count, but it's estimated that the brain contains around 85 billion glial cells.

Neurons (and glia) are not limited to the brain. They're found throughout the nervous system, which extends all over the body. One particularly large concentration can be found in the gut. The so-called enteric nervous system contains more neurons than the spinal cord, and around one-twentieth of the neurons of the brain. Moreover, and somewhat spookily, it can act independently of the rest of the nervous system. It would not be too much of a stretch to say that your gut harbours its own tiny brain. The second brain mostly concerns itself with digestion, but there is also evidence that signals from the gut play a large role in setting our moods.

Other myths about the brain persist, despite much evidence to the contrary. The statement 'We only use 10 per cent of our brains' is founded on precisely zero evidence. Would that mean we could remove 90 per cent of the brain and still function normally? It's also often said that we are born with all the brain cells we will ever need. As we grow, we do not acquire any new brain cells, but make more connections between those already there.

It is true that most nerve cells are produced in the womb. The lack of ability to make new cells partly explains why people rarely recover from paralysis after a serious injury to the spinal cord. There are exceptions, however. The brain continues to add neurons in the first few months after birth in a process called neurogenesis, while the hippocampus – an area strongly associated with the formation of memories – musters further cells throughout a person's life. And don't forget those glial cells. Unlike neurons, some types of glial cell are able to divide and repopulate.

An A-Z of
pseudoscience

It can be hard to separate the good science from the bogus.
Marketeers often use technical terms to make us buy their
product, safe in the knowledge that it has been rigorously tested.
Those who believe in the supernatural will sometimes use the
language of science to give fringe ideas a ring of credibility.
Here, then, is an alphabetical sample of some of the most widely
believed pseudoscience*.

Acupuncture This is the treatment of pain and other ailments by inserting needles into the body at specific points. Scientific tests of the practice have shown no benefits that can't be put down to the relaxing environment and placebo effect.

Biorhythms Can you predict a person's future well-being by analyzing the natural rhythms of their body? The theory of biorhythms posits three body cycles: a 23-day physical cycle, a 28-day emotional cycle and a 33-day intellectual cycle. The cycles begin at birth. By looking at how they intersect over time, we can work out a person's moods, strengths and maladies. In truth, there's no evidence such cycles exist (other than a tenuous link to the menstrual cycle in females), and the theory has zero scientific credibility.

Colonic irrigation Also known as colon cleansing or colon therapy, this treatment involves the heartwarming combination of a rectum, several metres of tubing and herb-supplemented water. The idea is to flush the large intestine of stubborn faecal matter, which can cling to the colon wall for months – much to the pleasure of parasites and bacteria. Happily, such cack-plaques are a myth. The benefits of colonic irrigation are anecdotal, and not supported by scientific study. In other words: a load of crap.

Detoxing No matter how many cucumber and acai berry smoothies you shove down your throat, you ain't going to detox. The idea that you can flush your body of impurities and toxins by drinking a detox drink makes no biological sense. A detox diet, high in fruit and vegetables, is probably good for you, but it won't rid your body of any lingering toxins that your kidney and liver aren't already handling. A triumph of marketing over reason.

Ear candles Ear candling, as you might imagine, involves sticking a lit candle in your ear. According to one manufacturer, the hollow candle 'stimulates the ear to eliminate the wax naturally', if inserting a fiery tube into your ear can be considered natural. The candles are also used to treat colds, tinnitus and headaches. Needless to say, this is another bogus therapy, with no proven benefit beyond a placebo effect.

Face on Mars The 1971 Viking 1 Mars probe snapped a convincing image of a giant face on the planet's surface. Many believed that this was evidence of an alien culture. Alas, later missions with sharper cameras destroyed the illusion. It's just a hill, which can look like a face if the Sun is shining in the right direction and you squint a bit. Our brains are instinctively inclined to look for patterns, particularly faces. The term pareidolia describes such cases, when we see things that aren't really there. The same psychological phenomenon may account for many ghost and UFO sightings.

Global warming denial The evidence is now overwhelming. It comes from dozens of separate lines of investigation and billions of datasets. The Earth's mean surface temperature is rising, and this is a consequence of human activities. And yet, many outside the scientific community still reject the fact that the world is warming, that humans have caused it, or both. Donald Trump, that dependable stalwart of the egregious example, once remarked that 'global warming was created by and for the Chinese in order to make US manufacturing non-competitive'. The science says otherwise. In a recent survey of research, only one author in more than 9,000 rejected the idea of human-led climate change. That's how strong the consensus is.

Homeopathy Homeopathy is based on the doctrine that like cures like. For example, you might take an extract of onion to cure a cold because onions are well known for causing tears and a runny nose. As if that concept weren't weird enough, the extract is diluted down to the point where the medicine is essentially pure water. Study after study has shown that homeopathic medicine works no better than a placebo. In other words, it's expensive nonsense that flies in the face of basic chemistry.

Intelligent design There are many definitions and interpretations of this phrase, but it basically boils down to 'God did it'. Adherents believe that life on Earth is so complex that it could only have been designed and fine-tuned by some creative force, not thrown up by the vagaries of natural selection. The notion is untestable, impossible to disprove and lacks any explanatory power over natural selection. Intelligent design is more religious belief than scientific theory.

Even so, it has occasionally found its way into the classroom as a rival theory to Darwinian evolution, particularly in the USA. By way of parody, opponents of ID have postulated their own creation story. According to them, an invisible and undetectable Flying Spaghetti Monster created the universe 'after drinking heavily'. It sounds crazy (and it is), but it's just as impossible to disprove, and just as useless at making sense of the world as traditional creationist doctrine. Bobby Henderson, a physics graduate who conceived of the Flying Spaghetti Monster, claims to have no beef with religion, only the teaching of religion as a science. 'I think we can all look forward to the time when these three theories are given equal time in our science classrooms across the country, and eventually the world,' he says; 'one third time for Intelligent Design, one third time for Flying Spaghetti Monsterism, and one third time for logical conjecture based on overwhelming observable evidence.'

Jesus on toast The Lord moveth in mysterious ways, none stranger than His habit of appearing in people's breakfasts. The image of Christ is so commonly seen in foodstuffs that Buzzfeed once ran an article called '22 People Who Found Jesus in Their Food'. These not-always-convincing cameos include Christ or his mother appearing on bananas, fish sticks and Marmite residue. It is, of course, impossible to rule out divine intervention, but a more likely explanation is to relegate such miracles to everyday pareidolia (see Face on Mars, above).

Kale The fibrous vegetable is the darling of dieticians, packed with vitamins, nutrients and antioxidants. It is surely king of the superfoods. Only, it's a fairy-tale kingdom. So-called superfoods are really just foods. Blueberries, quinoa, broccoli and their friends are all good and healthy, but are unlikely to have any miracle powers to warrant the 'super' prefix. Scientists might be able to say that a single compound extracted from goji berries can increase the lifespan of cells in a petri dish. It's a bold and unscientific leap to then assume that goji berries can slow down the ageing process in humans. (Also, they cost a fortune and taste like mildewed teabags.) Superfoods, kale included, are only worth buying if you genuinely prefer them over cheaper options.

Lie detectors Telling a lie can cause subtle changes to the body, such as increased heart rate or blushing. That much is true. But spotting such changes during interrogation is difficult and lie detectors are rarely used by the police. The best systems are more accurate than chance alone, but not by much (assuming that the many studies into this area are telling the truth). It is also possible to take countermeasures to beat a lie detector.

Moon landing conspiracy One of the more widely believed conspiracy theories is that Neil and Buzz never travelled any further than a film studio. Adherents point out numerous inconsistencies: shadows fall in multiple directions; the US flag appears to flutter despite the lack of a breeze; there are no stars in the sky. The long list of evidence for a hoax has been soundly debunked, and the Apollo landing sites have since been photographed from lunar orbit (unless they too are clever fakes, of course). One suspects it would have been far easier to fly to the Moon than to fake it, given the thousands of NASA employees and contractors who worked on the missions, and the intense media scrutiny of the time.

Numerology Numerology sounds like it might be a genuine branch of mathematics. It does deal with numbers and patterns, but in a hocus-pocus kind of way. Perhaps the most famous example in recent years is the bestselling *The Bible Code*, which picked out hidden messages in the Hebrew bible by selecting, say, every fiftieth letter from a given starting point. The idea and technique is easy to debunk. The message will quickly vanish if you use a slightly different source text, or even one with a typo. And given the huge number of possible starting points and intervals, you would expect to find meaningful combinations in any sufficiently long text. Indeed, famous 20th-century assassinations have been teased out of *Moby-Dick* and *War and Peace*. Either the technique is balderdash, or else Melville and Tolstoy should be venerated as prophets.

Out-of-body experiences A recent experiment shed new light on ghosts, doppelgangers and out-of-body experiences. Scientists were able to conjure up 'artificial spectres' by using shoulder-tapping robot limbs and clever feedback loops to mess with people's sense of their own bodies. The volunteers typically felt like someone was in the room with them. Illness or tiredness might play similar tricks on the mind, screwing with our perceptions of the body and its physical location in space. As Scrooge said to the ghost of Jacob Marley, employing the worst pun in the whole of literature: 'You may be an undigested bit of beef, a blot of mustard, a crumb of cheese, a fragment of underdone potato. There's more of gravy than grave about you'.

Perpetual motion 'Lisa, get in here. In this house we OBEY the laws of thermodynamics!' So admonished Homer Simpson after his precocious child built a perpetual motion machine. Lisa is not alone. Throughout history, humans have devised machines they think might run forever. They never do. Friction, heat loss or some other form of energy dissipation inevitably brings things to a halt. Even the planets, though apparently in perpetual rotation on a human timescale, will eventually come to a rest or be consumed by the Sun, which itself will one day burn out and evaporate.

Quantum nonsense The quantum world, as hinted elsewhere, is a very strange place indeed. Particles pop in and out of existence or hang around in opposite states at the same time. This very weirdness makes quantum mechanics a handy plaything for anyone who wants to explain away a crackpot theory without going to the trouble of doing any experiments. 'Telepathy is possible, and it works through quantum entanglement'; 'When we die, the soul leaves the body through quantum tunnelling'; 'Ghosts are a manifestation of wave-particle duality – they're just matter that has turned to waves' ... and so on. Be very wary of the word 'quantum' unless it's being used by someone qualified to wield it.

Remote viewing Is it possible to focus one's mind and witness events at a distant location? Imagine the applications. Everything from espionage to astronomy to perversion would undergo a revolution. Needless to say, plenty of money has been pumped into studying such parapsychological abilities. The book (and ploddy film) *The Men Who Stare at Goats* tells the story of a genuine US military programme that aims to find methods of spying or killing with the power of the mind. Sadly (or fortunately), no convincing evidence has ever been published, and nobody has a plausible mechanism (see quantum nonsense, above) for how such feats might work. Wishful thinking, in more than one sense.

Subliminal advertising 'Buy more popcorn'. 'Drink Fizz Cola!'. Can we be encouraged to buy a product by flashing a message so fleetingly that the conscious mind doesn't spot it? That's the idea behind subliminal advertising, a trick pioneered in the 1950s. The notion of selling to the subconscious feels a bit creepy. Indeed, it's been outlawed in the UK for over 50 years. And yet there's very little evidence to suggest subliminal advertising works. Scientific studies have found only the smallest of effects, and only in artificial, heavily controlled environments.

Telephone masts Many people would object to living beside a communications mast. They're not the most charming of erections, after all. But some say that masts present a more insidious problem than mere unsightliness. Numerous campaign groups have blamed the structures for a suite of ill effects, from nosebleeds to tumours. The claims have some resonance. Our towns are now bathed in an invisible nebula of microwaves, emanating from the masts on our roofs and the phones in our pockets. Surely it's feasible that this extra radiation might play havoc with our bodies? Well, it seems not. Study after study has found no link between increased exposure to microwaves and any ill health. Even common sense (though often unreliable) would agree. Microwave radiation is much less energetic than the visible light we all happily bask in. A small increase in exposure to less energetic microwaves is unlikely to have any ill effect. And if radiation emissions from phones *could* cause cancer, wouldn't we have seen a marked increase in people with tumours on their palms, thighs and ears?

UFOs In one sense, UFOs definitely do exist. The term is an abbreviation of unidentified flying object, which could be anything from an unusual species of bird to an unresolvable dot on the horizon. Whether alien spaceships visit the Earth is open to speculation, although it goes without saying that we lack any definitive and credible evidence. There is one thing that always puzzles me: how come UFOs are always linked to aliens? Wouldn't an equally plausible explanation be time-travelling humans, here from the future in advanced spaceships?

Vaccines lead to autism A 1998 study found a link between the measles-mumps-rubella (MMR) jab and autism. Thousands of alarmed parents refused to have their children inoculated, leading to outbreaks of mumps and measles. The research was later found to be fraudulent and no subsequent study (and there have been many) has found any evidence of a link. Unfortunately, the myth persists. Cases of the measles remain much higher than in the years before the vaccine scandal.

Water ionizers These devices are big business. The Amazon website lists dozens of competing products, including several that cost over £2,000. Water ionizers use an electric current to split tap water into hydrogen and oxygen. In doing so, water near the positive electrode becomes more alkaline. This is believed to be the good stuff, providing your body with pH balance, increased levels of oxygen, a boosted immune system, heaps more energy and even longer life. All of this is bunkum. Such devices do little more than basic water filtering, and there is no evidence to support any health benefits over drinking regular tap water. Many of the terms used by manufacturers, such as 'boost your immune system' and 'pH balance', are physiologically meaningless but sound impressive to the uncritical mind.

X causes cancer When not speculating about the royal family or getting excited by some pop star's new bikini look, the tabloid press is obsessed with cancer. Almost every day, we are told that some popular product or lifestyle choice can give you the disease or, conversely, ward it off. When the papers momentarily run out of substances that do or do not cause cancer, they instead uncover 'breakthrough' new research that will cure cancer within ten years. These stories usually have some grounding in science. A research team might have demonstrated, let us say, that a protein commonly found in dairy products can lead to an increased likelihood of tumour formation in mice, under laboratory conditions. Headline: Cheese Causes Cancer! To pick on the *Daily Mail* newspaper (it's an easy target), carcinogenic risks include candle-lit dinners, mouthwash, Facebook, being male and tap water. Such stories should usually be taken with a pinch of salt, though that too will double your risk of cancer.

Youthful looks Few sectors use technical-sounding jargon with such enthusiasm as the anti-ageing industry. Its adverts are full of unqualified phrases such as 'clinically proven' and 'dermatologist approved' – which sound impressive but are just vague enough to keep advertising watchdogs at bay. Baffling ingredients such as kojic acid and Boswelox further confound. There's a problem here: who tests this stuff? Cosmetics are unlikely to undergo the rigorous clinical trials you'd expect of a cancer drug. Usually, we only have the word of the manufacturer that the product is effective, or else the testimony of a celebrity, to whom the manufacturer is paying a large appearance fee. Where is the critical eye? Newspapers and magazines rarely intercede – 'New miracle skin cream' is a much better headline than 'This new product doesn't do very much', not to mention the huge advertising revenues that publishers make from the cosmetics industry. I'm not saying that all moisturizers and anti-wrinkle creams are useless, but we should turn a skeptical eye (with or without crow's feet) before spending silly money on a few grams of face goop.

Zoology (crypto-) Bit of a cheat for the letter Z. Cryptozoology is the search for animals not recognized by science, or else creatures believed to be long extinct. This includes such wonders as the unicorn, Bigfoot, the Loch Ness Monster and a surviving herd of woolly mammoths. This is an unusual branch of pseudoscience. While much here is nonsense, every once in a while a cryptid (as such creatures are known) really does come to light. Okapi and giant squid were considered myths before hard evidence emerged. Most famously, the coelacanth was known only through 65-million-year-old fossils before a living specimen was found in 1938. There are more things in heaven and earth than in the textbooks of scientists, but discoveries of new large animals are very rare.

---

* FOOTNOTE One might argue that presenting these phenomena as an A–Z list is itself pseudoscientific. Such a presentation makes it look like I've been pretty darned comprehensive ('Wow, he actually got something for X in there'), but instead it's a formula for cherry picking. Where, for example, is astrology? Why, other than for the sake of poop gags, have I chosen colonic irrigation over chiropractic or crop circles?

Famous scientists

Newton's apple, Darwin's finches and Einstein's bad maths grades.

$F = ma$

$F_1 = F_2$
$= G$

$F = ?$
$m \cdot a$

$\dfrac{m_1 \times m_2}{r^2}$

Newton developed his theory of gravity after a falling apple hit him on the head

A book such as this would be incomplete without tackling the most famous scientific myth of them all. Yet, surprisingly, there may be a pip of truth at the core of this one.

According to popular legend, the great mathematician was sitting in his mother's garden in Lincolnshire, in retreat from the plague of 1665–66. He chose a spot beneath an apple tree to contemplate the mysteries of the Universe. In particular, he pondered why the Moon should continuously circle the Earth.

His reverie was broken when a falling apple clobbered him on the periwig. The jolt sparked a eureka moment in which Newton saw that the apple and the Moon were attracted to the Earth by the same force: gravity. Both were falling but, unlike the apple, the Moon was also moving sideways, which balanced with the tug of gravity to ensure our satellite would never hit the Earth. He went on to derive his inverse-square law of gravitation and the laws of motion.

Well, possibly. Newton never wrote down his fruity encounter, and we are left only with the hearsay of others. Most persuasively from the pen of William Stukeley*, an antiquarian who befriended an aged Newton.

'After dinner, the weather being warm, we went into the garden & drank tea under the shade of some apple trees; only he & myself. Amid other discourse, he told me, he was just in the same situation, as when formerly the notion of gravitation came into his mind. Why sh[oul]d that apple always descend perpendicularly to the ground, thought he to himself; occasion'd by the fall of an apple, as he sat in contemplative mood ... If matter thus draws matter; it must be proportion of its quantity. Therefore the apple draws the Earth, as well as the Earth draws the apple.'

Newton related the anecdote to others, including Voltaire, so we can be certain that the story originates with him, and not some later writer. But did it ever happen? It's impossible to know. Newton was telling the tale at the very end of his life, 50 years after he began working on his theories. The story is likely to be embellished, or misremembered after all those decades. Plus, none of the sources mention that Newton was struck on the head. To call the tale apocryphal, however, is too harsh when its genesis lies with Newton himself.

The tree that supposedly inspired Newton can still be seen today. Visitors to Newton's birthplace of Woolsthorpe Manor, near Grantham, can view the ageing plant. Alas, you won't get to sit beneath it and contemplate the Universe. A willow screen was recently installed to prevent damage from inspiration-seeking tourists.

---

* FOOTNOTE This is the same William Stukeley who gave us the myth that the Great Wall of China can be seen from the Moon (see page 41).

Charles Darwin was the first to describe a theory of evolution

Here's a quote from Darwin. It's a bit wordy, but stick with it:

'Would it be too bold to imagine, that in the great length of time, since the Earth began to exist, perhaps millions of ages before the commencement of the history of mankind, would it be too bold to imagine, that all warm-blooded animals have arisen from one living filament, which THE GREAT FIRST CAUSE endued with animality, with the power of acquiring new parts, attended with new propensities, directed by irritations, sensations, volitions, and associations; and thus possessing the faculty of continuing to improve by its own inherent activity, and of delivering down those improvements by generation to its posterity, world without end!'

What Darwin is saying is that animals today are descended from very different creatures and, indeed, one common ancestor, which roamed the Earth millions of years ago. It's a Darwinian theory of evolution – though it comes not from Charles Darwin but his grandfather.

Erasmus Darwin* was a prodigious writer and thinker in his own right. He began speculating about evolution as early as 1789 – 70 years before his grandson would publish *On the Origin of Species*. He set out his ideas at some length, and even wrote poems about evolution, but had no clinching evidence to support his instinct.

It was radical thinking for the time. For most people, over most of history, the animals and plants of the Earth were changeless: a lion has always been a lion, and always would be. In the Christian tradition, God created the animals a day or two before he made humans. Similar creation stories are common throughout the world. The notion that animals might mutate into new forms would have seemed hilarious if not blasphemous to most people.

But this wasn't the only way of seeing the world, and nor was Erasmus the first to challenge it. Since ancient times, philosophers had courted the idea that new species might sometimes arise – though usually with the help of a divine hand. Erasmus Darwin put forward the most developed theory up to that point in his 1796 work *Zoonomia*. The idea was carried forward in the 19th century by the likes of Jean-Baptiste Lamarck, whose own flavour of evolution saw animals passing on traits that they'd acquired in life. Most famously, his theory explains the giraffe's long neck as the product of several generations, each striving a little harder to reach the leaves.

Robert Chambers's 1844 work *Vestiges of the Natural History of Creation* took things still further, placing the transmutation of species into the wider context of the cosmos. The book is flawed in many ways and contains numerous references to divine intervention. Darwin wasn't a fan of the content, but he did salute *Vestiges* for 'preparing the ground for the reception of analogous views' – that is, his own views, which were published a few years later as *On the Origin of Species*.

---

* FOOTNOTE Talented family, the Darwins. It's commonly noted that Charles married his cousin Emma Wedgwood, daughter of the famous potter Josiah. Less well known is that Charles and Emma's son George developed a leading theory for the Moon's formation (wrong, as it turned out), and would become the President of the Royal Astronomical Society. George's son Charles in turn worked out the fine structure of the hydrogen atom, and his daughter Gwen Raverat was a famous wood engraver. It doesn't stop there. Other relatives and descendants include economist John Maynard Keynes and the composer Ralph Vaughan Williams among many others.

And spare a thought for Patrick Matthew. This Scottish fruit farmer seems to have anticipated evolution by natural selection as early as 1831, when he published a convincing description of the mechanism. Unfortunately, he chose to do so in a book about naval timber, and then only as a note in the appendix. Nobody noticed until Matthew raised a public 'Ahem' when the *Origin* was published almost 30 years later. Darwin acknowledged Matthew's insight, writing: 'I think that no one will feel surprised that neither I, nor apparently any other naturalist, had heard of Mr Matthew's views, considering how briefly they are given'.

Rival theories continued to appear right up to the publication of *On the Origin Of Species* in 1859. Indeed, Darwin only published his long-mulled idea in that year after hearing that fellow naturalist Alfred Russel Wallace had independently arrived at similar conclusions. Darwin and Wallace jointly share the credit for the first publication to outline the idea – a paper presented to the Linnean Society on 1 July 1858. The *Origin* appeared 15 months later.

Oddly, some of the key phrases we associate with Darwin are scarcely found within the first edition of his greatest book. The word evolution is never used. Darwin employs the related term 'evolved' just once, and it is the final word in the book: '... from so simple a beginning endless forms most beautiful and most wonderful have been, and are being, evolved.' Nor did Darwin use the phrase 'survival of the fittest' on first publication; this was coined in 1864 by Herbert Spencer, after he had read Darwin's work. Darwin evidently liked this neat expression, for he incorporated it into later editions of the *Origin*.

The role of the finch in Darwin's great thesis is also overplayed. Many of us will remember this story from school. While travelling among the Galapagos, Darwin discovered that each island has its own species of finch, all well tailored to their environments. In a eureka moment, Darwin reasoned that an ancestor species must have flown over to the islands from the mainland and then, with time and geographical isolation, acquired a variety of new beak shapes to exploit different food sources. This contrasted with the traditional view that all species were created by God and could never change in form. The Galapagos finches therefore helped Darwin formulate his theory of evolution by natural selection. We might readily conjure up images of Darwin, sleeves rolled up, sitting on a Galapagos beach talking to a finch, wondering why its beak was particularly robust, while its cousin on the next island had a tiny pecker. It's doubtful any such encounter happened. Darwin seems to have ignored the finches of the island chain, focusing instead on mockingbirds. His expedition did capture a significant number of finches, and these would aid Darwin's later ponderings. But while on the Galapagos, Darwin left the task of collecting the finches (i.e. shooting the finches) to someone else, and did not consider their speciation until much later.

They weren't even finches, either. The specimens collected on the Galapagos are part of a family of birds known as tanagers, only distantly related to finches. Nevertheless, they look the part and the name has stuck. The term 'Darwin's Finches' was first coined in 1936 to refer collectively to the various island species.

Finches (or fake finches) are also largely absent from *On the Origin of Species*. The birds make just three appearances by name in the definitive sixth edition (although Darwin does discuss Galapagos birds generically). Ostriches, by contrast, pop their heads up 14 times. Thrushes make 18 visits, while pigeons steal the show with 113 mentions. The finch is more of a fixture (16 mentions) in *The Voyage of the Beagle*, Darwin's account of his round-the-world adventures, where he does make a few comments about the variation in the bird's beak.

Other oddities and misconceptions about the *Origin* abound. It's popularly supposed that the book deals with the rise of humans from monkeys and apes. It covers no such ground. Darwin would save that contentious subject for his 1871 book *The Descent of Man*. Even the title is open to confusion. The original and full appellation is *On the Origin of Species by Means of Natural Selection, or the Preservation of Favoured Races in the Struggle for Life*. It's usually abbreviated to *On the Origin of Species*. The sixth edition, however, drops the initial 'On'. I mention this seemingly trivial fact because I've often seen it written that it's wrong to call the book *The Origin of Species*. It's not wrong if we're referring to the sixth edition, often regarded as the definitive version. Book titles, like species, evolve.

As a student, Einstein was rubbish at maths

This mathematical maxim must have given succour to millions of underachievers over the years. If the greatest mind of the 20th century couldn't get his sums right, then there's hope for us all.

The story that Albert Einstein had failed his maths exams was popular during the physicist's own lifetime. It is easily refuted. There's good evidence that Einstein was top of his class even at primary school. He went on to excel in mathematics, striding well ahead of the curriculum and learning the rudiments of geometry and algebra by himself. In his own words, Einstein had 'mastered differential and integral calculus' before he was 15. Even today, many students will have no notion of calculus at that age.

Photographic reproductions of the school matriculation grades of a 17-year-old Einstein are readily available online. The certificate shows that the future Prof scored a maximum six points for algebra, geometry and physics, as well as history. Therein perhaps lies the source of the 'bad student' myth. Einstein attended school in Switzerland where the marking system reversed the more common German grading scheme, in which a score of one denoted excellence. To anyone brought up with the German system, a glance at Einstein's grades might suggest a struggling student. He also missed out, initially, on a place at the Swiss Federal Polytechnic in Zürich after failing to meet all the grades; his maths and physics marks were, however, outstanding.

It is true that Einstein had a few sweaty grapples with mathematics as he got older. His passion was always for physics, a subject not then so entrenched in complicated maths as it is today. He regularly skipped his maths classes, and later had to catch up when he needed more sophisticated equations to develop his theories. This was technical, high-level stuff. If Einstein was rubbish at maths because he took a little time over these techniques, then we might as well brand Pavarotti as a bad singer because he couldn't yodel under water.

It's often quipped that only a handful of people truly understand Einstein's theories of relativity. While that might once have been true (for the few weeks after the idea went public), it is no longer the case. Anyone taking degree-level physics should now grasp the fundamentals, and many scientists have an understanding deeper even than Einstein (as the field has moved on).

The hirsute professor questioned this myth himself in 1921, six years after his elucidation of general relativity. "It is absurd. Anyone who has had sufficient training in science can readily understand the theory. There is nothing amazing or mysterious about it. It is very simple to minds trained along that line, and there are many such in the United States."

DNA was discovered by Watson and Crick

Deoxyribonucleic acid, or DNA to its friends, is often called the blueprint of life. Its twisty ribbons are made up from millions of chemicals, whose order and patterns serve as a code. That code contains all the information needed to build and maintain an organism. This is true of all living species on Earth*, suggesting that animals, plants, bacteria and fungi all evolved from the same ancestor.

Nearly every cell in your body contains about 2 m (6½ ft) of DNA, coiled up into the cell nucleus – the equivalent of winding up a hosepipe the length of Europe into your garden shed, and still finding space for the lawnmower, workbench, stalled carpentry project and last year's Christmas tree.

The discovery of DNA is often attributed to James Watson and Francis Crick, working in Cambridge in 1953. In fact, DNA was discovered almost a century before. Swiss biologist Friedrich Miescher first isolated the substance in 1869 while playing around with the pus from hospital bandages. He wasn't entirely sure what he'd found, but since the extract came from the cell nucleus, he dubbed it 'nuclein'. German biochemist Albrecht Kossel dug much deeper a decade later. He was able to strip nuclein down to its constituent parts, removing proteins to isolate the more exotic components. He was the first to identify the building blocks of DNA, called nucleic acids. They come in four varieties that are commonly abbreviated to A, C, G and T, plus a fifth called U whose role isn't important for this discussion.

By 1919, scientists not only knew about DNA, they had also isolated its component parts, and more or less figured out how these parts were bonded together. The molecule's function remained a mystery, however, right up until the eve of Watson and Crick's work. Only in 1952 was it conclusively shown that DNA, not protein, is responsible for passing on genetic information to the next generation.

Watson and Crick's contribution – and it was one hell of a contribution – was to work out the three-dimensional structure of DNA. The double helix, whose rungs were built from A–T and C–G pairs, didn't just look pretty, it also looked functional. 'It has not escaped our notice,' remarked the authors in their oft-quoted write-up, 'that the specific pairing we have postulated immediately suggests a possible copying mechanism for the genetic material'. That mechanism was soon understood, and the field of molecular biology was born. Today, you can view a reconstruction of Watson and Crick's metallic model of the double helix in London's Science Museum.

Finally, it should be noted that Watson and Crick did not pull their structure out of thin air. Their insight rested squarely on the work of others, notably Rosalind Franklin, Maurice Wilkins and their co-workers at King's College London. Wilkins shared the 1962 Nobel Prize in Physiology or Medicine with the Cambridge duo. Franklin, alas, had died four years before.

---

* FOOTNOTE Many viruses, including ebola, HIV and SARS, are governed by RNA – a lengthy molecule similar to DNA. However, viruses are not usually considered life forms.

Are you saying it wrong?

Science is full of complicated, lengthy, technical words. Try saying 'acetylcholinesterase' after a few beers. Even the simpler phrases often carry contentious pronunciation. How would you fare with the following?

Betelgeuse The second brightest star in the constellation of Orion. Its peculiar name comes from an Arabic phrase meaning 'the hand of Orion'. Over the centuries, its spelling and pronunciation have shifted in many different ways. Today, it is most commonly pronounced as 'beetle-juice', just as in the 1988 Tim Burton movie. 'Bet-el-juice' is also well established.

Data Most people say 'day-ta'. A few pronounce it 'dah-ta'. Both can be considered correct, unless you're talking about the android commander from *Star Trek: The Next Generation*, when only 'day-ta' will do.

Euler Leonhard Euler was one of the great mathematicians of all time. It was he who first devised much of the notation found on any advanced mathematics course, including °, f(x), *e* and *i*. For present purposes (nitpicking and pedantry), you don't need to know what these symbols mean, only that Euler's name is pronounced 'oil-er' rather than 'yule-er'. It's a common error, repeated in the maths-inspired film *The Imitation Game*.

Large Hadron Collider Do not, I repeat, *do not* transpose the 'd' and the 'r' in that middle word.

Neanderthal Our ancient cousins pop up a number of times elsewhere in this book. As well as a history of mating with other species, Neanderthals are also promiscuous when it comes to their spelling and pronunciation. 'Neanderthal' is the original spelling, after the German valley in which fossilized remains were first discovered in 1856. The valley is now written as 'Neandertal' following a shift in German spelling, and the eponymous species of human is also commonly written without the 'h'. Both spellings are considered correct, though a strong argument could be made for 'Neanderthal', given that the Latin name of the species is fixed at *Homo neanderthalensis*. As for pronunciation, purists would plump for a hard 't' to match the original German, and this is the usual form among specialists. However, the 'th' sound is also common, especially in North America.

Patent Another of those words with transatlantic variation. Those brought up on British English tend to say 'pay-tunt', while those with American background usually favour 'pat-unt'.

Quark These subatomic particles should be rhymed with 'pork' rather than 'shark'. So reckons Murray Gell-Mann, one of the physicists who theorized quarks in 1964. Gell-Mann already had the phrase 'quork' rolling around his head. Then, by chance, he stumbled across the following passage in James Joyce's epically impenetrable book *Finnegans Wake*.

'Three quarks for Muster Mark!
Sure he has not got much of a bark
And sure any he has it's all beside the mark.'

Quarks tend to gather in threes, so the coincidence seemed apt. The trouble for Gell-Mann was that Joyce had clearly intended 'quark' to rhyme with 'mark' and 'bark'. But the physicist was insistent on the 'qwork' pronunciation (he argued that the 'three quarks for Muster Mark' were really 'three quarts' – a measure of beer – and so his pronunciation could be justified). Most physicists today will say 'qwork', a name reinforced by the Ferengi bartender in *Star Trek: Deep Space Nine*, but 'qwark' is also common.

Uranus For obvious reasons, the professionals tend to pronounce this one as 'yura-nus' rather than 'your-anus'. The most overused pun in astronomy almost didn't happen. The planet's discoverer, William Herschel, wanted to call it George's Star, after the reigning British monarch.

Let's start a new wave of false facts

Having debunked many scientific myths, we need to replace them with some fresh bilge.

- Isaac Newton would dye his hair a different colour of the rainbow every week. Few people realised, as he rarely went out in public, and always wore a wig.

- Sound cannot propagate through outer space, except for the B-flat semitone. Nobody knows why.

- Earth's Moon might not be made of cheese, but Saturn's moon Hyperion is formed from a substance similar to tofu.

- Homo *dentificans* is a lost species of human which hunted in the wastelands of the north during the last Ice Age. These hominids were distinguished by their thick, woolly hides, and tusk-like incisors.

- Charles Darwin's luxuriant beard contained a thriving colony of bullfinches.

- Ada Lovelace was not only the world's first computer programmer, but she is also credited with inventing the lolcat and rickrolling.

- Element number 67 is named Holmium in honour of Sherlock Holmes. It is a clever pun on the detective's fondness for the phrase 'elementary, my dear Watson'.

- The Nobel Prize for Chemistry was not awarded in 1952, due to a general shortage of interesting science.

- It really is possible to slow down time by holding your breath. The effect is tiny, though, and could only be measured with complex equipment.

- When swearing his or her allegiance to the Crown, the Astronomer Royal must promise never to utter jokes about Uranus in a public forum.

- Although he never wore one in life, Michael Faraday was buried in a lab coat.

- The original plans for the Large Hadron Collider included provision for a 'scientifically themed ghost train' that would run through the tunnels in a bid to boost public engagement.

- Numbats are the only known animals to have triple-stranded DNA.

Further information

The scientific literature is vast, truly vast. One estimate from 2010 suggests that 50 million scholarly articles have been published since the Royal Society first started the trend 350 years ago. A new one appears every 20 seconds or so. Add to that the countless textbooks, magazine articles, websites, popular science books, videos and apps, and the scope for further learning is approximately infinite. For that reason, any kind of bibliography must be hopelessly, laughably incomplete.

Instead, I offer only a handful of places to start. One of the best 'beginner's guides' of recent years must be Bill Bryson's *A Short History of Nearly Everything* (Doubleday, 2005). A non-scientist himself, Bryson brings a refreshing outsider's perspective to the big questions of the Universe, and is particularly strong at introducing the various personalities from the history of science. I also enjoyed Yuval Noah Harari's *Sapiens* (Harper, 2014), for the history of our species. For physics and cosmology, nobody can ever quite beat the various books by Carl Sagan for inspiring wonder in the simplest language. Everybody should read Darwin's *On the Origin of Species* – it is one of the great masterpieces of human thought and, unlike Newton's *Principia*, is easy and pleasurable to digest.

If you live in a big town or city, look out for meet-up groups by a community who call themselves skeptics (always with a 'k', contrary to British spelling). These are gatherings for people with enquiring, critical minds, keen to debunk and overturn various forms of pseudoscience. These meet-ups are always in an informal setting such as a bar or pub, and usually include an invited speaker of some reputation. Not only will you learn plenty of science, you'll also sharpen your critical thinking, while making a new set of friends. Have a listen to the excellent *The Skeptics' Guide to the Universe* podcast to get a flavour of this growing community.

Science grows, evolves and makes new discoveries all the time, so it's important to keep up with new developments. Most reputable newspapers now do an excellent job of covering important research in some depth. To go still further, try buying a dedicated magazine, such as *New Scientist* or *Scientific American*. You don't have to have a scientific background to enjoy such periodicals – they're pitched at the curious non-scientist as much as those who work in the sector.

So follow your nose, seek out whatever interests you and, above all, keep nitpicking.

Acknowledgements

With thanks, as ever, to my wife Heather, who tolerated my frequent absences to write this book during the first few months of our daughter Holly's life.

About the author

Matt Brown holds degrees in Chemistry (BSc) and Biomolecular Science (MRes). He has served as a scientific editor and writer at both Reed Elsevier and Nature Publishing Group, and has contributed to two previous science books. He served as the Royal Institution's quizmaster for several years, and has also put on science quizzes for the Royal Society, Manchester Science Museum, STEMPRA and the Hunterian Museum. He gave a lecture about Michael Faraday while rotating around the London Eye. With comedian Helen Keen, he co-hosted a series of successful 'Spacetacular!' science-themed stage shows, culminating in a sell-out 2013 show at Leicester Square Theatre.

Matt is also the author of *London Night and Day* (2015) and *Everything You Know About London is Wrong* (2016), both from Batsford. He also serves as Editor-at-large of Londonist.com.

Contact him on i.am.mattbrown@gmail.com with any notes, questions, queries, nitpicks or offers of beer.

Index